1983

A pathway into number theory

R. P. BURN

A pathway into number theory

CAMBRIDGE UNIVERSITY PRESS
Cambridge
London New York New Rochelle
Melbourne Sydney

Published by the Press Syndicate of the University of Cambridge
The Pitt Building, Trumpington Street, Cambridge CB2 1RP
32 East 57th Street, New York, NY 10022, USA
296 Beaconsfield Parade, Middle Park, Melbourne 3206, Australia

First published 1982

Printed in Great Britain at the
University Press, Cambridge

Library of Congress catalogue card number: 81 10013

British Library Cataloguing in Publication Data

Burn, R. P.
A pathway into number theory.

1. Numbers, Theory of
I. Title
512'.7 QA241

ISBN 0 521 24118 9 hard covers
ISBN 0 521 28534 8 paperback

CONTENTS

INTRODUCTION

The construction of the *Pathway*

Have you attended a mathematics lecture, followed each step of the argument, and yet at the end felt that you did not understand what it was about? Have you read a proof of a theorem in a book and felt the same? If so, you have experienced a feeling common to most mathematicians.

This book on number theory has been put together by keeping a record of how I actually resolved the blocks which I encountered as I read a number of standard texts. Time and again, it was the exploration of special cases which illuminated the generalities for me. This collection of explorations was then organised into a sequence in such a way that the 'pathway' would climb towards the standard theorems which occur here as problems for the student at the end of each section.

The motivation for assembling the *Pathway* was a college need to mount a course for which lectures would not be given. If the *Pathway* is more successful than some other books or undergraduate lecture courses in number theory, it is because it follows more closely than usual the natural process of discovery, and puts logic in its proper place. 'The purpose of rigour', said Hadamard, 'is to legitimate the conquests of the intuition, and it has never had any other purpose'. Formality, abstraction and generality have an essential place in the completion of any piece of mathematics, but their role in discovery is varied. In the *Pathway*, the introduction to each new idea is as informal and as specific as I could make it. There are a few remarks about foundations in the notes, but the statement of the Peano axioms post-dates almost all of the number theory in the book and is not given here.

In the selection of material, the needs of future teachers have been kept in mind. The theme of sums of squares links chapters 4, 5, 6, 8,

9 and 10. Arithmetic functions and the distribution of primes were thought to offer less connection with school work.

In the seventeenth century, Fermat complained that the mathematics of his time was so dominated by geometry that numbers in their own right had not received their due attention. Many exponents of number theory, in the spirit of Fermat, have been jealous of the autonomy of their subject and it was only late in the nineteenth century that Minkowski made mathematicians once more aware of the rich interplay of number and space. There is a taste of Minkowski's work in chapter 9, and I have taken every opportunity to exploit the inter-relation of geometry and number and to plan the display of numerical information so as to facilitate the visual perception of number patterns. (The exception to this is in chapter 10 because continued fractions have already been explored geometrically at this level, in the textbook of Stark (1978).)

The interdependence of chapters is indicated by arrows here.

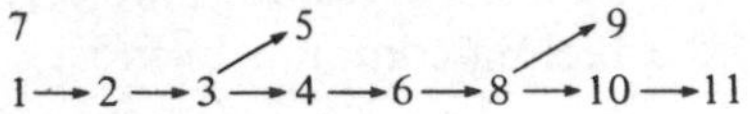

Advice to the student

The sequence of questions and notes in this book will lead you through an undergraduate course in number theory. You need a pocket calculator with a square-root function, a reciprocal function, and preferably two memories. Because of the concrete approach adopted here, there are many questions in each chapter which may be tackled without reference to earlier chapters and it is only rarely that a question presupposes detailed knowledge of work in a preceding chapter. The notes on each chapter contain solutions, comments and sometimes definitions, and they are meant to be read.

If you are intending to pursue this 'pathway' from start to finish, you should have some familiarity with complex numbers, mathematical induction, 2×2 matrices and groups up to Lagrange's theorem, as these topics appear in a modern sixth-form course such as the Advanced level books of the *School Mathematics Project* (Cambridge University Press 1979). Apart from this background, the *Pathway* is self-contained, although those who have pursued a first course in analysis will have a richer appreciation of the convergence of continued fractions and the appeal to the mean value theorem in the proof of Liouville's theorem than would be appropriate to develop here.

The bibliography has been annotated to encourage concurrent rather than further reading.

In the text, a reference to 'q 31' refers to question 31 in the same chapter. A reference to 'q 2.31' refers to question 31 of chapter 2, and a reference to 'n 2.31' refers to the note on question 2.31. At the foot of each page the reference '⟦x⟧' indicates that the notes for the questions on that page appear on page x.

Acknowledgements

I am happy to acknowledge my debt to those who have shaped in me the direction and method of construction of this book; to Alan Bell who showed me that the sequence of definition followed by theorem followed by exercise is reversed in the normal process of learning, and to Bill Brookes who pointed out the value of recording one's mathematical 'blocks' and the processes of removing them.

I am grateful to Dr S. M. Edmonds for holding me to my belief that a course consisting of problems could be put together; to Professor J. F. Adams for suggesting number theory as an appropriate field; to Dr S. J. Patterson for suggesting the major theme of sums of squares, for indicating into how many avenues this would lead and for the proof of Legendre's theorem in chapter 9; to Dr A. F. Beardon for the proof that the index of a subgroup of a group of isometries of a lattice is equal to the measure of the fundamental unit of the tessellation in chapter 9; to Dick Tahta for graphical displays of $\sum_{d|12} \phi(d)$ and of Pythagorean triangles; to Dr C. W. L. Garner for providing an excellent working environment at Carleton University, Ottawa, while work on the typescript was progressing in the spring of 1980; to my colleagues at Homerton for their interest as the *Pathway* took shape, and most particularly to Stuart Plunkett for his analysis of the various ways in which information about numbers may be visually displayed, for a computer programme which let me watch quadratic forms take shape on a graphics terminal and for the numerical display of $\sum_{d|12} \phi(d)$; and finally to those students at Homerton whose investigations into number theory I have been allowed to share, and particularly Jane Charman, whose work has shaped parts of chapters 3 and 6.

Homerton College, Cambridge R. P. Burn
January 1981

1

The fundamental theorem of arithmetic

Division algorithm

1 Look at table 1.1. If the same pattern was extended downwards, would it eventually incorporate any positive integer $\{1, 2, 3, \ldots, n, n+1, \ldots\}$ that we might care to name?

2 What is the relation between each number in table 1.1 and the number below it?

3 Give a succinct description of the full set of numbers in the column below 0.

4 If you choose two numbers from the column below 0 and add them together, where in the table must their sum lie?

5 The whole of the array in table 1.1 may be considered as an addition table with the column below 0 down one side and the numbers 1, 2, 3 across the top. Using your brief description of the numbers in the column below 0, devise a comparably succinct description of the full set of numbers in the column below 1, and similarly succinct descriptions of the sets of numbers in the other two columns.

6 If two numbers lie in the second column and the lesser is subtracted from the greater, where does the difference lie?

7 If two numbers lie in the third column and the lesser is subtracted from the greater, where does the difference lie? Try to prove your result in a general way which would apply to all such pairs.

8 If two numbers lie in the fourth column and the lesser is subtracted from the greater, where does the difference lie? Prove it.

Table 1.1

0	1	2	3
4	5	6	7
8	9	10	11
12	13	14	15
16	17	18	19
20	21	22	23
24	25	26	27
28	29	30	31
32	33	34	35
36	37	38	39
40	41	42	43
44	45	46	47
48	49	50	51
52	53	54	55
56	57	58	59
60	61	62	63
64	65	66	67
68	69	70	71
72	73	74	75
76	77	78	79
80	81	82	83
84	85	86	87
88	89	90	91
92	93	94	95
96	97	98	99
100	101	102	103
104	105	106	107
108	109	110	111
112	113	114	115
116	117	118	119
120	121	122	123
124	125	126	127
128	129	130	131
132	133	134	135
136	137	138	139
140	141	142	143
144	145	146	147
148	149	150	151
152	153	154	155
156	157	158	159
160	161	162	163
164	165	166	167
168	169	170	171
172	173	174	175
176	177	178	179
180	181	182	183
184	185	186	187
188	189	190	191
192	193	194	195
196	197	198	199

9 If two numbers are chosen, both from the second column, where does their sum lie? Prove your claim generally.

10 If two numbers are chosen, both from the fourth column in table 1.1, where does their sum lie? Prove your claim generally.

11 Are there general rules which enable you to fill in the table below for addition of numbers by columns? If only the numbers at the heads of the columns are used in this table, the table that results is an example of an *addition* table *modulo* 4. Such a table is denoted by $(\mathbf{Z}_4, +)$.

+	Number in column 0	Number in column 1	Number in column 2	Number in column 3
Number in column 0				
Number in column 1				
Number in column 2				
Number in column 3				

12 Two numbers which lie in the same column of table 1.1 are said to be *congruent modulo* 4. We write $5 \equiv 13 \pmod 4$. Give an algebraic definition of $a \equiv b \pmod 4$.

13 Is it true that every positive integer is expressible in exactly one of the forms $4q$, $4q+1$, $4q+2$, $4q+3$ for some integer $q \geqslant 0$? How would you determine which of the four types a particular number, say 1553, might be?

14 Investigate the effect of multiplication on the columns of table 1.1. Is it possible to construct a multiplication table analogous to the one above for addition?

⟦18⟧

Table 1.2

0	1	2	3	4
5	6	7	8	9
10	11	12	13	14
15	16	17	18	19
20	21	22	23	24
25	26	27	28	29
30	31	32	33	34
35	36	37	38	39
40	41	42	43	44
45	46	47	48	49
50	51	52	53	54
55	56	57	58	59
60	61	62	63	64
65	66	67	68	69
70	71	72	73	74
75	76	77	78	79
80	81	82	83	84
85	86	87	88	89
90	91	92	93	94
95	96	97	98	99
100	101	102	103	104
105	106	107	108	109
110	111	112	113	114
115	116	117	118	119
120	121	122	123	124
125	126	127	128	129
130	131	132	133	134
135	136	137	138	139
140	141	142	143	144
145	146	147	148	149
150	151	152	153	154
155	156	157	158	159
160	161	162	163	164
165	166	167	168	169
170	171	172	173	174
175	176	177	178	179
180	181	182	183	184
185	186	187	188	189
190	191	192	193	194
195	196	197	198	199

15 In table 1.2, five columns have been used to display the positive integers. Give a general description for the set of numbers in each of the columns.

16 Is it true that every positive integer is expressible in exactly one of the forms $5q$, $5q+1$, $5q+2$, $5q+3$, $5q+4$ for some integer $q \geqslant 0$? How would you determine which of the five types a particular number, say 6666, might be?

17 Make addition and multiplication tables modulo 5. Give a formal justification of at least two entries in each table.

18 Table 1.1 displayed the positive integers in four columns, and table 1.2 displayed the positive integers in five columns. Generally, if the positive integers (with zero) are displayed in b columns, what are the numbers in the first row, and what are the numbers in the first column? Can every positive integer be expressed as the sum of two integers, one in the first row and one in the first column? If the integer a appears in the same row as the integer bq, what is the relationship between the three numbers bq, $b(q+1)$ and a? Deduce that $a = bq + r$, where $r = 0$ or r is a positive integer $< b$.

19 If a and b are positive integers, and q_1, q_2, r_1, r_2 are positive integers or zero, with r_1 and $r_2 < b$, and if moreover, $a = bq_1 + r_1 = bq_2 + r_2$, prove that the difference between r_1 and r_2 is a multiple of b, and deduce that $r_1 = r_2$ and $q_1 = q_2$.

(Questions 18 and 19 together give the *division algorithm.*)

20 In table 1.3, the columns have been extended both upwards and downwards from the row 0, 1, 2, 3. Give general descriptions of the set of numbers in each column. Does the pattern for the addition of columns found in q 11 still hold with the columns extended upwards? Does the pattern for the multiplication of columns found in q 14 still hold with the columns extended upwards?

21 To which column of table 1.3 (if extended) would -161 belong?

22 In the group of integers under addition $(\mathbf{Z}, +)$, the subset in the first column of table 1.3 is denoted by $4\mathbf{Z}$, and in the other columns by $4\mathbf{Z}+1$, $4\mathbf{Z}+2$ and $4\mathbf{Z}+3$ respectively. Describe these four subsets of $(\mathbf{Z}, +)$ using the language of group theory.

23 Propose a form of the division algorithm which would apply to any integer a and any positive integer b.

⟦19⟧

Table 1.3

−100	−99	−98	−97
−96	−95	−94	−93
−92	−91	−90	−89
−88	−87	−86	−85
−84	−83	−82	−81
−80	−79	−78	−77
−76	−75	−74	−73
−72	−71	−70	−69
−68	−67	−66	−65
−64	−63	−62	−61
−60	−59	−58	−57
−56	−55	−54	−53
−52	−51	−50	−49
−48	−47	−46	−45
−44	−43	−42	−41
−40	−39	−38	−37
−36	−35	−34	−33
−32	−31	−30	−29
−28	−27	−26	−25
−24	−23	−22	−21
−20	−19	−18	−17
−16	−15	−14	−13
−12	−11	−10	−9
−8	−7	−6	−5
−4	−3	−2	−1
0	1	2	3
4	5	6	7
8	9	10	11
12	13	14	15
16	17	18	19
20	21	22	23
24	25	26	27
28	29	30	31
32	33	34	35
36	37	38	39
40	41	42	43
44	45	46	47
48	49	50	51
52	53	54	55
56	57	58	59
60	61	62	63
64	65	66	67
68	69	70	71
72	73	74	75
76	77	78	79
80	81	82	83
84	85	86	87
88	8	90	91
92	93	94	95
96	97	98	99

24 Justify the form of the division algorithm which you have proposed.

The division algorithm, almost as it stands, is the basis of congruence arithmetic. We shall use it to prove the fundamental facts about the factorisation of natural numbers in the rest of this chapter. In number systems other than **N**, an analogue of the division algorithm can sometimes be proved, and, in such systems, a unique factorisation theorem follows. See for example, q 5.53.

Greatest common divisor and Euclidean algorithm

25 For the two chains of number sets given here, devise a rule for moving along an arrow, and a rule for when to stop.

$$\{57, 36\} \to \{21, 36\} \to \{21, 15\} \to \{6, 15\} \to \{6, 9\}$$
$$\to \{6, 3\} \to \{3, 3\} = \{3\} \text{ stop.}$$

$$\{98, 175\} \to \{98, 77\} \to \{21, 77\} \to \{21, 56\} \to \{21, 35\}$$
$$\to \{21, 14\} \to \{7, 14\} \to \{7, 7\} = \{7\} \text{ stop.}$$

Construct a sequence of number sets using the same pattern as that given above, starting with {170, 130}.

26 If $a > b$, what is the successor to $\{a, b\}$ according to the pattern of chains in q 25? If the chain starts from two positive integers, why can no negative integer, or zero, appear in the chain?

27 Use the first chain of number sets given in q 25 to find a pair of integers x, y for each of the following equations.

$57 = 57x + 36y$,
$36 = 57x + 36y$,
$21 = 57x + 36y$,
$15 = 57x + 36y$,
$6 = 57x + 36y$,
$9 = 57x + 36y$,
$3 = 57x + 36y$.

28 Use the second chain of number sets given in q 25 to express each of the numbers 175, 98, 77, 21, 56, 35, 14, 7 in the form $98x + 175y$, where x and y are integers.

29 Does the set of numbers $\{57x + 36y \mid x, y \in \mathbf{Z}\}$ form a subgroup of $(\mathbf{Z}, +)$? What is the smallest positive number in this set? Must every multiple of this number be in the set? Must every number in the set be a multiple of this number?

30 Suggest a simple description of the subgroup of $(\mathbf{Z}, +)$ generated by the numbers 57 and 36.

⟦19⟧

31 Does the set of numbers $\{98x+175y \mid x, y \in \mathbf{Z}\}$ form a subgroup of $(\mathbf{Z}, +)$? What is the smallest positive number in this set? Must every multiple of this number be in the set? Must every number in the set be a multiple of this number?

32 Suggest a simple description of the subgroup of $(\mathbf{Z}, +)$ generated by the numbers 98 and 175.

33 Give a formal description of the subgroup of $(\mathbf{Z}, +)$ generated by the non-zero integers a and b.
If d is the smallest positive integer in this subgroup, explain why every multiple of d lies in the subgroup.
If the subgroup were to contain a number c which was not a multiple of d, use the division algorithm to prove that the subgroup would have to contain a positive integer smaller than d. This contradiction establishes that the subgroup of $(\mathbf{Z}, +)$ generated by a and b is in fact generated by the single number d.

34 If a and b are non-zero integers and the subgroup of $(\mathbf{Z}, +)$ which they generate together is generated by the single positive integer d, explain why

$d \mid a$ and $d \mid b$

(d divides a and d divides b, or d is a factor of a and d is a factor of b), and by returning to the original description of the group generated by a and b, prove that $d = ax + by$ for some integers x and y, and deduce that every factor which is common to a and b, divides d. This makes d the highest common factor, or *greatest common divisor* of a and b, and we write $d = \gcd(a, b)$.

35 Use q 34 to explain why there must exist integers x and y such that $2x+3y=1$. Find by experiment a pair of integers which satisfy this equation.
If $2a+3b=1$ and $2c+3d=1$, prove that $2(a-c)=3(d-b)=6t$ for some integer t and deduce that every solution of $2x+3y=1$ has the form

$x=-4+3t, \quad y=3-2t.$

36 Prove that the set of integers $\{12x+18y+27z \mid x, y, z \in \mathbf{Z}\}$ forms a subgroup of $(\mathbf{Z}, +)$. Prove that every element of this subgroup has a factor 3. By an appropriate choice of x, y and z, prove that 3 is an element of this subgroup. Deduce that the subgroup is the cyclic group generated by 3 and that $\gcd(12, 18, 27)=3$.

37 State and prove an analogue of q 33 for subgroups of $(\mathbf{Z}, +)$ generated by three non-zero integers.

⟦20⟧

38 State and prove an analogue of q 34 for the greatest common divisor of three non-zero integers.

39 The chains of q 25 for finding the greatest common divisor of two numbers are usually abbreviated as follows

$\{57, 36\} \to \{36, 21\} \to \{21, 15\} \to \{15, 6\} \to \{6, 3\}$ stop

and

$\{175, 98\} \to \{98, 77\} \to \{77, 21\} \to \{21, 14\} \to \{14, 7\}$ stop,

and in this form are referred to as the *Euclidean algorithm.* The larger number is written first and the process stops when the smaller number is a factor of the larger number. Determine which steps in q 25 have been omitted in the Euclidean algorithm.

40 Use a pocket calculator with two memories, or two calculators, or pencil and paper, to find gcd (107 360, 30 866).

41 Use the division algorithm to describe the steps of the Euclidean algorithm. Explain why each pair in the chain have the same gcd.

42 Is it possible for two adjacent terms in the Fibonacci sequence 1, 1, 2, 3, 5, 8, 13, . . . , where each term is the sum of its two predecessors, to have a gcd different from 1?

Unique factorisation into primes

43 A positive integer p, different from 1, is called a *prime number* when its only positive factors are 1 and p. If p is prime and n is a positive integer, what values can gcd (p, n) have?

44 Since $p|a$ implies $p \leqslant a$, 2 is a prime number. Since 3 is not a multiple of 2, 3 is a prime number. Since 5 is not a multiple of 2, 3 or 4, 5 is a prime number. List the prime numbers less than 30.

45 In the table below, by using tracing paper if you prefer, delete all the numbers except 2 which are multiples of 2; delete all the numbers except 3 which are multiples of 3; delete all the numbers except 5 which are multiples of 5; delete all the numbers except 7 which are multiples of 7.
Are all those numbers which remain, prime numbers?

1	2	3	4	5	6	7	8	9	10
11	12	13	14	15	16	17	18	19	20
21	22	23	24	25	26	27	28	29	30
31	32	33	34	35	36	37	38	39	40
41	42	43	44	45	46	47	48	49	50
51	52	53	54	55	56	57	58	59	60
61	62	63	64	65	66	67	68	69	70
71	72	73	74	75	76	77	78	79	80
81	82	83	84	85	86	87	88	89	90
91	92	93	94	95	96	97	98	99	100

⟦20⟧

Table 1.4

2	4	6	8
10	12	14	16
18	20	22	24
26	28	30	32
34	36	38	40
42	44	46	48
50	52	54	56
58	60	62	64
66	68	70	72
74	76	78	80
82	84	86	88
90	92	94	96
98	100	102	104
106	108	110	112
114	116	118	120
122	124	126	128
130	132	134	136
138	140	142	144
146	148	150	152
154	156	158	160
162	164	166	168
170	172	174	176
178	180	182	184
186	188	190	192
194	196	198	200

Table 1.5

1	5	9	13	17	21	25	29	33	37
41	45	49	53	57	61	65	69	73	77
81	85	89	93	97	101	105	109	113	117
121	125	129	133	137	141	145	149	153	157
161	165	169	173	177	181	185	189	193	197
201	205	209	213	217	221	225	229	233	237
241	245	249	253	257	261	265	269	273	277
281	285	289	293	297	301	305	309	313	317
321	325	329	333	337	341	345	349	353	357
361	365	369	373	377	381	385	389	393	397
401	405	409	412	417	421	425	429	433	437
441	445	449	453	457	461	465	469	473	477
481	485	489	493	497	501	505	509	513	517
521	525	529	533	537	541	545	549	553	557
561	565	569	573	577	581	585	589	593	597
601	605	609	613	617	621	625	629	633	637
641	645	649	653	657	661	665	669	673	677
681	685	689	693	697	701	705	709	713	717
721	725	729	733	737	741	745	749	753	757
761	765	769	773	777	781	785	789	793	797
801	805	809	813	817	821	825	829	833	837
841	845	849	853	857	861	865	869	873	877
881	885	889	893	897	901	905	909	913	917
921	925	929	933	937	941	945	949	953	957
961	965	969	973	977	981	985	989	993	997

46 Prove that any non-prime number between 1 and 101 has 2, 3, 5 or 7 as a factor. Suppose $a = bc$ where neither b nor c is 1, and deal separately with the two cases $b = c$ and $b > c$.

47 The procedure used in q 45 is called *Eratosthenes' sieve*. What prime numbers must be used to determine in this way all the prime numbers up to 1000?

48 The numbers 3, 5, 7 and 11 are all prime numbers and $5 \cdot 7 - 3 \cdot 11 = 2$. Find another set of four prime numbers p, q, r and s such that $pq - rs = 2$. Can you find four prime numbers p, q, r and s such that $pq - rs = 1$? Would you expect to be able to find four distinct prime numbers p, q, r and s such that $pq - rs = 0$?

49 Why must the product of two even numbers always be even? All the numbers in table 1.4 are even numbers. Delete those numbers which are products of even numbers from the table. Those numbers which remain are, in a sense, 'prime' within this set. Using this convention, determine precisely which numbers are 'prime' and which are not. Express the number 180 as a product of 'prime' numbers in two different ways. Find another even number which may be expressed as a product of 'primes' in two different ways.

50 Why must the product of two numbers of the form $4n + 1$ again be of this form?
All the numbers in table 1.5 are of the form $4n + 1$. Delete all those numbers which are products of numbers in the table different from 1. Those numbers different from 1 which remain are, in a sense, 'prime' within this set. Using this convention, determine precisely which numbers, up to 101, are 'prime' and which are not. Express the number 441 as a product of 'prime' numbers in two different ways. Find one other number in table 1.5 which may be expressed as a product of 'prime' numbers in two different ways.

51 In ordinary arithmetic we presume that if a prime number p divides a product ab, then p divides one or other of the factors. Is this presumption valid within the even numbers with their own 'prime' convention, or within the set of numbers of the form $4n + 1$ with their own 'prime' convention?

52 If p, q and r are prime numbers and $p \neq q$, what is gcd (p, q), and what is the smallest positive integer of the form $px + qy$, where x and y are integers? What is the smallest positive integer of the form $rpx + rqy$? If, moreover, $p | rq$, prove that $p | r$ and deduce that $p = r$.

⟦21⟧

53 Use the result of q 52 to prove that it is not possible to find four distinct prime numbers p, q, r and s such that $ps-qr=0$.

54 Generalise the argument of q 52 to prove that if p is a prime number, and a and b are non-zero integers such that $p|ab$, then either $p|a$ or $p|b$. (Suppose that p does not divide a.)

55 Using the result of q 54, prove that it is not possible to find five distinct prime numbers p, q, r, s, and t such that $pq=rst$, and indeed that it is not possible to find five prime numbers satisfying this equation, distinct or not.

56 If p, q, r and s are prime numbers, prove that $p|qr$ implies that $p=q$ or $p=r$, and that $p|qrs$ implies that $p=q$ or r or s.

57 If $p_1, \ldots, p_n$ and $q_1, \ldots, q_m$ are prime numbers, not necessarily all distinct, and if

$$p_1p_2 \ldots p_n = q_1q_2 \ldots q_m$$

prove that p_1 is equal to at least one of the q_i. After dividing through by this factor, prove that p_2 is equal to another of the q_i. By an induction on the number of factors prove that $n=m$ and that each of the p_i equals one of the q_i.
Express 5 247 000 as a product of prime numbers.

58 If $p_1, \ldots, p_n$ are the distinct prime factors of a positive integer a greater than 1, prove that

$$a = p_1^{\alpha_1}p_2^{\alpha_2} \ldots p_n^{\alpha_n},$$

where the α_i are unique positive integers.
(*Fundamental theorem of arithmetic*)

59 Using their factorisation into primes, find the gcd and lcm of 5 247 000 and 189 280.
By extending the convention of exponents to permit an exponent of 0, devise a formula for the greatest common divisor of two numbers, and also a formula for the least common multiple of two numbers.
Prove that the product of two numbers is equal to the product of their gcd and lcm.

60 If gcd $(a, b)=1$, the integers a and b are said to be *coprime*. If gcd $(a, b)=1$ and gcd $(a, c)=1$, prove that gcd $(a, bc)=1$.

61 If a and b are coprime and $a|bc$, prove that $a|c$.

⟦21⟧

Infinity of primes

62 The number $2 \cdot 3 \cdot 5 \cdot 7 \cdot 11 \cdot 13+1$ is not prime. What can you say about its prime factors?

63 If p and q are any two prime numbers, what can you say about the prime factors of $pq+1$?

64 Suppose the prime numbers are listed in order: $p_1=2$, $p_2=3$, $p_3=5$, $p_4=7$, $p_5=11$, etc. Suppose further that

$p_m | p_1p_2 \ldots p_n+1.$

What can you say about m?

Deduce that the total number of primes cannot be finite and that

$p_{n+1} \leqslant p_1p_2 \ldots p_n+1.$

65 If a is a positive integer, prove that every prime factor of $4a-1$ is odd.

Which columns of table 1.1 contain prime numbers?

If a product of four odd prime numbers lies in the right hand column of table 1.1, how many of the four prime numbers must lie in that column?

If a product of n odd prime numbers is congruent to 3 modulo 4, how many of the n prime numbers must be congruent to 3 modulo 4?

What can you say about the prime factors of

$4p_2p_3p_4 \ldots p_n-1$?

Deduce that the total number of primes congruent to 3 modulo 4 cannot be finite.

66 Examine the factors of numbers of the form $3a-1$, and prove that the total number of primes congruent to 2 modulo 3 cannot be finite.

Mersenne primes

67 Use a calculator to write down the numbers 2^n-1 for $n=1, \ldots, 16$.

If $2^n-1=m$, does your list provide evidence for either of the propositions

if n is prime, then m is prime

or

if m is prime, then n is prime?

(When m is prime it is called a *Mersenne prime*.)

⟦22⟧

Notes and answers

For preliminary reading, see bibliography, especially Reid (1956), Ore (1967) and Butts (1973). For concurrent reading see also Davenport (1968).

1 Number theory is principally about the natural numbers

$\mathbf{N}=\{1, 2, 3, \ldots, n, \ldots\}$

but often it is best to consider these numbers in a wider context, such as the integers

$\mathbf{Z}=\{0, \pm 1, \pm 2, \ldots, \pm n, \ldots\}$,

the rational numbers

$\mathbf{Q}=\left\{\frac{p}{q}\,\middle|\, p, q \in \mathbf{Z}, q \neq 0\right\}$,

the real numbers **R**, which correspond to the set of all points on a straight line, or the complex numbers

$\mathbf{C}=\{x+\mathrm{i}y \mid x, y \in \mathbf{R}, \mathrm{i}^2=-1\}$.

2 Four less or four more.

3 Multiples of 4.

4 $4n+4m=4(n+m)$.

5 $4n+1$, $4n+2$, $4n+3$.

6 $(4n+1)-(4m+1)=4(n-m)$ in the first column.

7 $(4n+2)-(4m+2)=4(n-m)$.

8 $(4n+3)-(4m+3)=4(n-m)$.

9 $(4n+1)+(4m+1)=4(n+m)+2$ in the third column.

10 $(4n+3)+(4m+3)=4(n+m+1)+2$ in the third column.

11

0	1	2	3
1	2	3	0
2	3	0	1
3	0	1	2

12 $a-b$ has a factor 4.

13 $4\underline{|1\,553}$

388 remainder 1, so $1553=4\cdot 388+1$.

14

0	0	0	0
0	1	2	3
0	2	0	2
0	3	2	1

For example,

$$(4n+2)(4m+2)$$
$$=4(4mn+2m+2n+1).$$

15 $5n$, $5n+1$, $5n+2$, $5n+3$, $5n+4$.

16 $5\,\underline{|6666}$

1333 remainder 1, so $6666 = 5 \cdot 1333 + 1$.

17 Addition

0	1	2	3	4
1	2	3	4	0
2	3	4	0	1
3	4	0	1	2
4	0	1	2	3

Multiplication

0	0	0	0	0
0	1	2	3	4
0	2	4	1	3
0	3	1	4	2
0	4	3	2	1

$(5n+2)+(5m+3) = 5(n+m+1)$.

$(5n+2)(5m+4) = 5(5mn+2m+4n+1)+3$.

18 First row $0, 1, 2, \ldots, b-1$.
First column $0, b, b \cdot 2, b \cdot 3, \ldots$
$bq \leqslant a < b(q+1)$, so $0 \leqslant a - bq < b$, and $0 \leqslant r < b$.
It is a fundamental property of the natural numbers that by repeatedly adding 1, we can exceed any given number. The argument here is based on the assumption that by repeatedly adding b we can exceed a, and $b(q+1)$ is the first such multiple.
The letter q stands for quotient and the letter r for the remainder on dividing a by b.

19 $b(q_1 - q_2) = r_2 - r_1$, but $0 \leqslant r_1, r_2 < b$, so $-b < r_1 - r_2 < b$. But $r_2 - r_1$ is a multiple of b, so $r_2 - r_1 = 0$.

20 $4n$, $4n+1$, $4n+2$, $4n+3$. Yes. Yes.

21 $-161 = 4\,(-41)+3$.

22 $4\mathbf{Z}$ is a subgroup of $(\mathbf{Z}, +)$ and $4\mathbf{Z}+1$, $4\mathbf{Z}+2$, $4\mathbf{Z}+3$ are its cosets.

23 Given any integer a and a positive integer b, there is a unique integer q and a unique integer r with $0 \leqslant r < b$, such that $a = bq + r$.

24 Choose q so that $a - bq$ is the smallest possible positive integer or zero. Then $a - b(q+1) < 0 \leqslant a - bq$. Thus $a - bq < b$ and $0 \leqslant r = a - bq < b$. The proof of uniqueness follows the one for q 19.

25 $\{170, 130\} \to \{40, 130\} \to \{40, 90\} \to \{40, 50\} \to \{40, 10\}$
$\to \{30, 10\} \to \{20, 10\} \to \{10, 10\} = \{10\}$ stop.

26 $\{a, b\} \to \{a-b, b\}$. If $a > b > 0$, then $b > 0$ and $a - b > 0$. If $a - b = 0$, then $a = b$ and $\{a, b\} = \{a\}$.

27 $(1, 0)$, $(0, 1)$, $(1, -1)$, $(-1, 2)$, $(2, -3)$, $(-3, 5)$, $(-5, 8)$.

28 $(0, 1)$, $(1, 0)$, $(-1, 1)$, $(2, -1)$, $(-3, 2)$, $(-5, 3)$, $(-7, 4)$, $(9, -5)$.

29 From q 27, 3 is in the set; $57x + 36y = 3(19x + 12y)$.

30 Multiples of 3.

31 From q 28, 7 is in the set; $98x+175y=7(14x+25y)$.

32 Multiples of 7.

33 $\{ax+by \mid x, y \in \mathbf{Z}\}$. If $d=ax+by$, then $nd=a(nx)+b(ny)$. If $c \neq dq$ for any q, then $c=dq+r$ for $0<r<d$. But c in the subgroup implies $c-dq$ in the subgroup.

34 The subgroup generated by d consists of all its multiples, so a and b are multiples of d and $d|a$, $d|b$. Since d is in the subgroup generated by a and b, $d=ax+by$ for some integers x and y. Now any factor common to a and b divides $ax+by$ and so divides d.

If you have not come across the result before, you will be surprised by the usefulness of the fact that $\gcd(a, b)=ax+by$ for some integers x and y. We use it in our proof of the uniqueness of factorisation into primes, and at some stage in most of the chapters of this book.

35 $\gcd(2, 3)=1$, so there are integers x and y such that $2x+3y=1$, for example $2(-4)+3\cdot 3=1$.
$2(a-c)=3(d-b)$ implies that $d-b$ is an even integer so $3(d-b)=6t$.
Now $a=c+3t$, $b=d-2t$ and the result follows since $c=-4$, $d=3$ is one solution.

36 Check closure, identity, inverses; $12x+18y+27z=3(4x+6y+9z)$
$12(-2)+18\cdot 0+27\cdot 1=3$.

37 If d is the least positive integer in the set $\{ax+by+cz \mid x, y, z \in \mathbf{Z}\}$, every element in the subgroup is a multiple of d by the argument of n 33. Since the subgroup is closed, it contains every multiple of d.

38 Take d as in n 37, then the argument there proves that $d|a$, $d|b$, $d|c$. If $d=ax+by+cz$, then every common divisor of a, b and c divides d.

39 If $a>b$, and a is not a multiple of b, then $\{a, b\} \to \{b, a-bq\}$ where $0< a-bq<b$.

40 1342.

41 $57=36+21$, $175=98+77$,
$36=21+15$, $98=77+21$,
$21=15+6$, $77=21\cdot 3+14$,
$15=6\cdot 2+3$, $21=14+7$,
$6=3\cdot 2$, $14=7\cdot 2$.
If $a>b$ and a is not a multiple of b, then $\{a, b\} \to \{b, a-bq\}$.
If $d=\gcd(a, b)$, then $d|b$ and $d|a-bq$.
If $c=\gcd(b, a-bq)$, then $c|(a-bq)+bq=a$.
So $c|a$ and $c|b$, so $c|d$.

42 If $a_{n+1} = a_n + a_{n-1}$, then any factor common to a_{n+1} and a_n is also a factor of a_{n-1}. By repeating this argument, any such factor must be a factor of all preceding terms of the Fibonacci sequence.

43 gcd $(p, n) = 1$ or p.

44 We presume a is positive. 2, 3, 5, 7, 11, 13, 17, 19, 23, 29.

45 They should be.

46 $10^2 = 100$, so if $a = b^2 < 101$, $b \leqslant 10$, and so b has a prime factor less than 10, and thus a does. If $a = bc < 101$ with $b > c$, then $b > c > 10$ is impossible, so $c < 10$, and c and a have a prime factor < 10.

47 $31^2 = 961$, $32^2 = 1024$. It is sufficient to consider the primes up to 31. If you have opportunity, run a computer programme which will print the prime numbers up to 1000.

48 The uniqueness of factorisation into primes is something that is taken for granted in school arithmetic. This question and the next two try to expose the need for a proof.
$3 \cdot 19 - 5 \cdot 11 = 2$.
$3 \cdot 5 - 2 \cdot 7 = 1$.

49 Multiples of 4 are not 'prime'.
Numbers of the form $4n + 2$ are 'prime'.
$180 = 18 \cdot 10 = 6 \cdot 30$.
$60 = 6 \cdot 10 = 2 \cdot 30$.

50 $(4m + 1)(4n + 1) = 4(4mn + m + n) + 1$.
5, 9, 13, 17, 21, 29, 33, 37, 41, 49, 53, 57, 61, 69, 73, 77, 89, 93, 97 are 'primes'.
$441 = 21 \cdot 21 = 9 \cdot 49$.
$693 = 21 \cdot 33 = 9 \cdot 77$.

51 No.

52 gcd $(p, q) = 1$, so there are integers x and y such that $px + qy = 1$, and $rpx + rqy = r$.
$p|rq$ implies $p|rpx + rqy = r$. But r is a prime, so $p = r$.

53 $ps = qr$ implies $p|qr$, so from q 52 either $p = q$ or $p = r$.

54 If p does not divide a, then gcd $(p, a) = 1$, so there are integers x and y such that $px + ay = 1$, and $bpx + bay = b$, so $p|bpx + bay = b$.

55 $pq = rst$ implies $p|rst$, so $p|r$ or $p|st$ by q 54. Now by q 52 $p = r$ or s or t. If we take $p = r$, we have $q = st$ which contradicts the fact that q is a prime.

56 $p|qr$ implies $p = q$ or $p = r$ by q 52.
$p|qrs$ implies $p|qr$ or $p|s$ by q 54, so $p = q$ or r or s.

57 $p_1 | p_1 p_2 \ldots p_n = q_1 q_2 \ldots q_m$ so $p_1 | q_1$ or $p_1 | q_2 \ldots q_m$ by q 54.
If $p_1 \neq q_1$, then $p_1 | q_2$ or $p_1 | q_3 \ldots q_m$.
If $p_1 \neq q_1, q_2$, then $p_1 | q_3$ or $p_1 | q_4 \ldots q_m$, etc.
So $p_1 = q_1$ or q_2 or q_3 or $\ldots$ or q_m.
Suppose $p_1 = q_1$, then $p_2 \ldots p_n = q_2 \ldots q_m$.
If prime factorisation is presumed unique for $n-1$ factors the argument given shows that it is unique for n factors. Prime factorisation is trivially unique for $n = 1$, so it follows generally by induction.
$5\,247\,000 = 2^3 \cdot 3^2 \cdot 5^3 \cdot 11 \cdot 53$

58 If $p_i | a$, let α_i be the largest positive integer such that $p_i^{\alpha_i} | a$. Then $p_1^{\alpha_1} p_2^{\alpha_2} \ldots p_n^{\alpha_n} | a$.
So $a = k p_1^{\alpha_1} p_2^{\alpha_2} \ldots p_n^{\alpha_n}$. Now the only prime factors of a are $p_1, \ldots, p_n$ so by q 57 the only prime factors of k are $p_1 \ldots p_n$. But the α_i are maximal, so no $p_i | k$. So $k = 1$.

59 $189\,280 = 2^5 \cdot 5 \cdot 7 \cdot 13^2$, so gcd $= 2^3 \cdot 5$ and lcm $=$ $2^5 \cdot 3^2 \cdot 5^3 \cdot 7 \cdot 11 \cdot 13^2 \cdot 53$.
If $a = p_1^{\alpha_1} p_2^{\alpha_2} \ldots p_n^{\alpha_n}$ and $b = p_1^{\beta_1} p_2^{\beta_2} \ldots p_n^{\beta_n}$, where $p_1, \ldots, p_n$ include all the prime factors of a or b, then

gcd $(a, b) =$ product of all $p_i^{\min\{\alpha_i, \beta_i\}}$

lcm $(a, b) =$ product of all $p_i^{\max\{\alpha_i, \beta_i\}}$

Now $\min\{\alpha, \beta\} + \max\{\alpha, \beta\} = \alpha + \beta$, whichever is the greater. So [gcd (a, b)][lcm (a, b)] $= ab$.

60 *Either* gcd $(a, b) = 1$ implies $ax + by = 1$ for some x, y, so $cax + cby = c$. If gcd $(a, bc) = d$, $d | c$. Since gcd $(a, c) = 1$, $d = 1$.
Or consider prime factors of a, b and c.

61 gcd $(a, b) = 1$ implies that there exist integers x and y such that $ax + by = 1$. Now $acx + bcy = c$, so $a | bc$ implies $a | c$.

62 Prime factors $\neq 2$, 3, 5, 7, 11, 13.

63 Prime factors $\neq p$, q.

64 $m \neq 1, 2, \ldots, n$ so $m > n$.
This construction provides a prime number outside any finite set of prime numbers, so the complete set of primes is not finite.
There is a prime $> p_n$ which divides $p_1 \ldots p_n + 1$ so p_{n+1} is equal to or less than this prime.
So $p_{n+1} \leqslant p_1 \ldots p_n + 1$.

65 Even $\times$ even $=$ even.
Even $\times$ odd $=$ even.
So $4a - 1$ which is odd, has no factor 2.
Every odd prime either has the form $4n + 1$ or the form $4n + 3$.

Modulo 4

×	1	3
1	1	3
3	3	1

If a product of four odd primes $= 4n+3$, either one or three of them must be of the form $4n+3$. $4p_2p_3 \ldots p_n - 1$ is odd and is of the form $4n+3$. Thus it has an odd number of prime factors (and hence at least one) of the form $4n+3$. But all prime factors of this number are greater than p_n, so we can construct a prime number of this form greater than any given prime of this form.

66 $3a-1$ is congruent to 2 modulo 3.

Modulo 3

×	1	2
1	1	2
2	2	1

so it must contain an odd number of factors congruent to 2 modulo 3. But $2 \cdot 3 \cdot p_3p_4 \ldots p_n - 1$ is not divisible by any prime up to p_n so all its prime factors are greater, and so we can construct a prime number of the form $3n+2$ which is greater than any given prime of this form.

67 $2^{ab}-1=(2^a-1)(2^{a(b-1)}+2^{a(b-2)}+\ldots+2^{a\cdot 2}+2^a+1)$.
So if m is prime then n is prime. $2^{11}-1=23 \cdot 89$. Mersenne's interest in these numbers stems from the fact that when 2^p-1 is a prime, the sum of the divisors of $2^{p-1}(2^p-1)$ is equal to $2^{p-1}(2^p-1)$.

Historical note

The theorems in this chapter are essentially from Euclid (c. 300 B.C.); explicitly so in the case of the division algorithm and the algorithm for finding the highest common factor of two given numbers. Euclid's notation precluded a complete statement or proof either of the fundamental theorem of arithmetic or of the claim that there exist an infinity of primes. The essential arguments of these theorems are his, though he only discussed products of up to three or four primes. C. F. Gauss (in 1801) gave the first explicit statement and proof of the fundamental theorem in the general form in which we know it today, although mathematicians before Gauss had assumed the result. In 1775, L. Euler claimed that every arithmetic progression beginning with 1 contained infinitely many primes. In 1837, Dirichlet proved that if a and b are coprime, then $\{an+b \mid n \in \mathbf{N}\}$ contains infinitely many primes.

2

Modular addition and Euler's ϕ function

Congruence classes and the Chinese remainder theorem

1 If a and b are numbers in table 2.1 and you are given that 6 is a factor of $a-b$, what can you say about the positions of a and b in the table?

2 If 6 is a factor of $a-b$, we write

$$a \equiv b \pmod 6$$

and say that a is congruent to b modulo 6.
Explain why $a \equiv a \pmod 6$ for all integers a.
If $a \equiv b \pmod 6$, prove that $b \equiv a \pmod 6$.
If $a \equiv b \pmod 6$ and $b \equiv c \pmod 6$, prove that $a \equiv c \pmod 6$.

3 Determine the set of all integers congruent to 0 (mod 6).
Determine the set of all integers congruent to 1 (mod 6).
Determine the set of all integers congruent to 2 (mod 6).
Determine the set of all integers congruent to 3 (mod 6).
Determine the set of all integers congruent to 4 (mod 6).
Determine the set of all integers congruent to 5 (mod 6).
These six sets are called the *congruence classes* or *residue classes* modulo 6. Use the division algorithm to prove that each integer belongs to exactly one of these classes.

4 When n is a factor of $a-b$, we write

$$a \equiv b \pmod n$$

and say that *a is congruent to b modulo n*.
We presume here that a, b and n are integers, and n is positive.
State and prove generalised forms of q 2 and q 3 for integers modulo n.

Table 2.1

−90	−89	−88	−87	−86	−85
−84	−83	−82	−81	−80	−79
−78	−77	−76	−75	−74	−73
−72	−71	−70	−69	−68	−67
−66	−65	−64	−63	−62	−61
−60	−59	−58	−57	−56	−55
−54	−53	−52	−51	−50	−49
−48	−47	−46	−45	−44	−43
−42	−41	−40	−39	−38	−37
−36	−35	−34	−33	−32	−31
−30	−29	−28	−27	−26	−25
−24	−23	−22	−21	−20	−19
−18	−17	−16	−15	−14	−13
−12	−11	−10	−9	−8	−7
−6	−5	−4	−3	−2	−1
0	1	2	3	4	5
6	7	8	9	10	11
12	13	14	15	16	17
18	19	20	21	22	23
24	25	26	27	28	29
30	31	32	33	34	35
36	37	38	39	40	41
42	43	44	45	46	47
48	49	50	51	52	53
54	55	56	57	58	59
60	61	62	63	64	65
66	67	68	69	70	71
72	73	74	75	76	77
78	79	80	81	82	83
84	85	86	87	88	89
90	91	92	93	94	95

5 By using tracing paper or by making a copy of the table, enter the numbers 0, 1, 2, 3, 4 and 5 in the appropriate square of the array given below. For example, $4 \equiv 0 \pmod 2$ and $4 \equiv 1 \pmod 3$.

Modulo 2 \ Modulo 3	0	1	2
0			
1			

6 By using tracing paper or by making a copy of the table, enter the numbers 0, 1, 2, . . . , 9, 10, 11 in the appropriate square of the array given below.

Modulo 3 \ Modulo 4	0	1	2	3
0				
1				
2				

7 By using tracing paper or by making a copy of the table, enter the numbers 0, 1, 2, . . . , 33, 34, 35 in the appropriate square of the array given below.

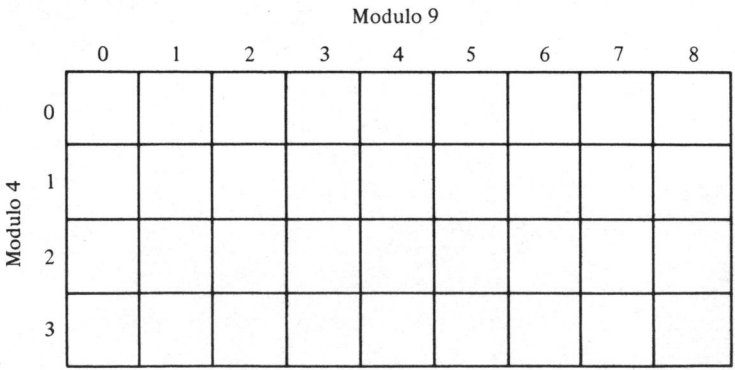

Modulo 4 \ Modulo 9	0	1	2	3	4	5	6	7	8
0									
1									
2									
3									

⟦40⟧

8 $432 = 16 \times 27$.
Imagine entering the numbers 0, 1, 2, . . . , 429, 430, 431 in a 16×27 array with each column containing numbers congruent modulo 27 and each row containing numbers congruent modulo 16. Can you be sure that no two of the integers between 0 and 431 should be entered in the same position? Must there be exactly one of the numbers between 0 and 431 in each square?

9 By using tracing paper or by making a copy of the table, enter the numbers 0, 1, 2, . . . , 9, 10, 11 in the appropriate square of the array given below.

Modulo 2 \ Modulo 6	0	1	2	3	4	5
0						
1						

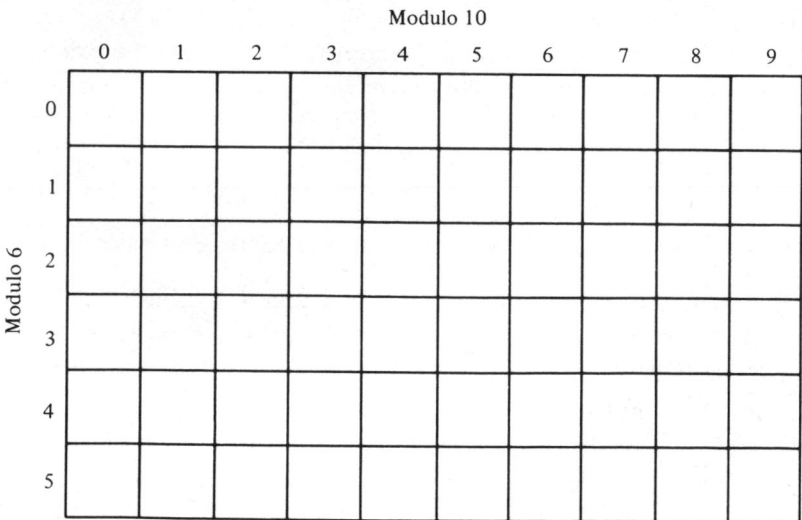

Why is reasoning analogous to that used in q 8 not available here?

10 By using tracing paper or by making a copy of the table, enter the numbers 0, 1, 2, . . . , 57, 58, 59 in the appropriate square of the array given below.

Modulo 6 \ Modulo 10	0	1	2	3	4	5	6	7	8	9
0										
1										
2										
3										
4										
5										

Why is the reasoning used in q 8 not applicable here?

⟦40⟧

11 If you were to generalise the argument of q 8 to establish that there are not two distinct numbers between 0 and $mn-1$ which are congruent modulo m and congruent modulo n, what conditions should be imposed on m and n to avoid the situation in q 9 and q 10?

12 If a and b are two integers such that

$$a \equiv b \pmod m$$

and

$$a \equiv b \pmod n,$$

prove that $a \equiv b \pmod{mn}$ if m and n are coprime. Deduce that if $0 \leq a, b \leq mn-1$, then $a = b$.

13 How can you be sure that there is a unique integer x, $0 \leq x \leq mn-1$, such that

$$x \equiv a_1 \pmod m$$

and

$$x \equiv a_2 \pmod n,$$

when m and n are coprime?

14 Find x, $0 \leq x < 15$, when $x \equiv 2 \pmod 3$ and $x \equiv 4 \pmod 5$.

15 Find x, $0 \leq x < 120$, when $x \equiv 14 \pmod{15}$ and $x \equiv 5 \pmod 8$.

16 Find x, $0 \leq x < 120$, when $x \equiv 2 \pmod 3$, $x \equiv 5 \pmod 8$ and $x \equiv 4 \pmod 5$.

17 Extend the arguments of q 12 and q 13 to establish the existence of a unique solution modulo $m_1m_2m_3$ to

$$x \equiv a_1 \pmod{m_1},$$
$$x \equiv a_2 \pmod{m_2},$$
$$x \equiv a_3 \pmod{m_3},$$

provided that m_1, m_2, m_3 are relatively coprime in pairs.

18 Extend the argument of q 17 by induction to establish the existence of a unique solution modulo $m_1m_2 \ldots m_n$ to the n congruences

$$x \equiv a_i \pmod{m_i}, \qquad i = 1, 2, \ldots, n,$$

where $m_1, m_2, \ldots, m_n$ are relatively coprime in pairs. (*Chinese remainder theorem*)

This theorem completes our study of the intersection of congruence classes. We shall use the Chinese remainder theorem to prove that Euler's ϕ function is multiplicative in 50. Now we begin to consider operations on congruence classes.

The groups ($\mathbf{Z}_n$, +) and their generators

19 Are the sets $\{6n+1 | n \in \mathbf{Z}\}$ and $\{6n-5 | n \in \mathbf{Z}\}$ identical?

20 Explain why each integer is congruent to exactly one of the integers −12, −5, 8, −3, −8, 23, modulo 6. Such a set is called a *complete set of residues* modulo 6 because each residue class modulo 6 is represented once.

21 If $a_1, a_2, a_3, a_4, a_5, a_6$ are six integers with no two of them congruent modulo 6, must they form a complete set of residues modulo 6?

22 If $a \equiv b \pmod 6$ and $c \equiv d \pmod 6$, does it follow that $a+c \equiv b+d \pmod 6$ and $a-c \equiv b-d \pmod 6$?

23 Use the complete set of residues 0, 1, 2, 3, 4, 5 to make an addition table modulo 6 (see q 1.11).
Is it a group table? What is the identity? Is there a single element which generates the group?

24 Use the complete set of residues 12, −5, 8, −3, −8, 23 to make an addition table modulo 6. Does the structure match that of q 23? What is the identity? Is there a single element which generates the group?

25 If $\{a_1, a_2, a_3, a_4, a_5, a_6\}$ is any complete set of residues modulo 6, can you be sure that a group table can be made using the elements of this set and writing the sum of a_i and a_j as a_k where $a_i + a_j \equiv a_k \pmod 6$. This form of addition on a complete set of residues is called addition modulo 6. Must this table exhibit the same structure as that of q 23? What is the identity? Is there a generator?

26 If $a \equiv b \pmod n$ and $c \equiv d \pmod n$, prove that $a+c \equiv b+d \pmod n$ and $a-c \equiv b-d \pmod n$.

27 Use the examples of the groups ($\mathbf{Z}_3$, +), . . . , ($\mathbf{Z}_{10}$, +) exhibited in table 2.2 to determine the first ten terms of the sequences

1, 1+1, 1+1+1, . . .

2, 2+2, 2+2+2, . . .

3, 3+3, 3+3+3, . . .

under addition modulo 3, 4, 5, 6, 7, 8, 9, 10.

⟦41⟧

Table 2.2

Addition modulo 3

0	1	2
1	2	0
2	0	1

Addition modulo 4

0	1	2	3
1	2	3	0
2	3	0	1
3	0	1	2

Addition modulo 5

0	1	2	3	4
1	2	3	4	0
2	3	4	0	1
3	4	0	1	2
4	0	1	2	3

Addition modulo 6

0	1	2	3	4	5
1	2	3	4	5	0
2	3	4	5	0	1
3	4	5	0	1	2
4	5	0	1	2	3
5	0	1	2	3	4

Addition modulo 7

0	1	2	3	4	5	6
1	2	3	4	5	6	0
2	3	4	5	6	0	1
3	4	5	6	0	1	2
4	5	6	0	1	2	3
5	6	0	1	2	3	4
6	0	1	2	3	4	5

Addition modulo 8

0	1	2	3	4	5	6	7
1	2	3	4	5	6	7	0
2	3	4	5	6	7	0	1
3	4	5	6	7	0	1	2
4	5	6	7	0	1	2	3
5	6	7	0	1	2	3	4
6	7	0	1	2	3	4	5
7	0	1	2	3	4	5	6

Addition modulo 9

0	1	2	3	4	5	6	7	8
1	2	3	4	5	6	7	8	0
2	3	4	5	6	7	8	0	1
3	4	5	6	7	8	0	1	2
4	5	6	7	8	0	1	2	3
5	6	7	8	0	1	2	3	4
6	7	8	0	1	2	3	4	5
7	8	0	1	2	3	4	5	6
8	0	1	2	3	4	5	6	7

Table 2.2 (*cont.*)

Addition modulo 10

0	1	2	3	4	5	6	7	8	9
1	2	3	4	5	6	7	8	9	0
2	3	4	5	6	7	8	9	0	1
3	4	5	6	7	8	9	0	1	2
4	5	6	7	8	9	0	1	2	3
5	6	7	8	9	0	1	2	3	4
6	7	8	9	0	1	2	3	4	5
7	8	9	0	1	2	3	4	5	6
8	9	0	1	2	3	4	5	6	7
9	0	1	2	3	4	5	6	7	8

Addition modulo 11

0	1	2	3	4	5	6	7	8	9	10
1	2	3	4	5	6	7	8	9	10	0
2	3	4	5	6	7	8	9	10	0	1
3	4	5	6	7	8	9	10	0	1	2
4	5	6	7	8	9	10	0	1	2	3
5	6	7	8	9	10	0	1	2	3	4
6	7	8	9	10	0	1	2	3	4	5
7	8	9	10	0	1	2	3	4	5	6
8	9	10	0	1	2	3	4	5	6	7
9	10	0	1	2	3	4	5	6	7	8
10	0	1	2	3	4	5	6	7	8	9

Addition modulo 12

0	1	2	3	4	5	6	7	8	9	10	11
1	2	3	4	5	6	7	8	9	10	11	0
2	3	4	5	6	7	8	9	10	11	0	1
3	4	5	6	7	8	9	10	11	0	1	2
4	5	6	7	8	9	10	11	0	1	2	3
5	6	7	8	9	10	11	0	1	2	3	4
6	7	8	9	10	11	0	1	2	3	4	5
7	8	9	10	11	0	1	2	3	4	5	6
8	9	10	11	0	1	2	3	4	5	6	7
9	10	11	0	1	2	3	4	5	6	7	8
10	11	0	1	2	3	4	5	6	7	8	9
11	0	1	2	3	4	5	6	7	8	9	10

Addition modulo 13

0	1	2	3	4	5	6	7	8	9	10	11	12
1	2	3	4	5	6	7	8	9	10	11	12	0
2	3	4	5	6	7	8	9	10	11	12	0	1
3	4	5	6	7	8	9	10	11	12	0	1	2
4	5	6	7	8	9	10	11	12	0	1	2	3
5	6	7	8	9	10	11	12	0	1	2	3	4
6	7	8	9	10	11	12	0	1	2	3	4	5
7	8	9	10	11	12	0	1	2	3	4	5	6
8	9	10	11	12	0	1	2	3	4	5	6	7
9	10	11	12	0	1	2	3	4	5	6	7	8
10	11	12	0	1	2	3	4	5	6	7	8	9
11	12	0	1	2	3	4	5	6	7	8	9	10
12	0	1	2	3	4	5	6	7	8	9	10	11

Table 2.2 (*cont.*)

Addition modulo 14

0	1	2	3	4	5	6	7	8	9	10	11	12	13
1	2	3	4	5	6	7	8	9	10	11	12	13	0
2	3	4	5	6	7	8	9	10	11	12	13	0	1
3	4	5	6	7	8	9	10	11	12	13	0	1	2
4	5	6	7	8	9	10	11	12	13	0	1	2	3
5	6	7	8	9	10	11	12	13	0	1	2	3	4
6	7	8	9	10	11	12	13	0	1	2	3	4	5
7	8	9	10	11	12	13	0	1	2	3	4	5	6
8	9	10	11	12	13	0	1	2	3	4	5	6	7
9	10	11	12	13	0	1	2	3	4	5	6	7	8
10	11	12	13	0	1	2	3	4	5	6	7	8	9
11	12	13	0	1	2	3	4	5	6	7	8	9	10
12	13	0	1	2	3	4	5	6	7	8	9	10	11
13	0	1	2	3	4	5	6	7	8	9	10	11	12

Addition modulo 15

0	1	2	3	4	5	6	7	8	9	10	11	12	13	14
1	2	3	4	5	6	7	8	9	10	11	12	13	14	0
2	3	4	5	6	7	8	9	10	11	12	13	14	0	1
3	4	5	6	7	8	9	10	11	12	13	14	0	1	2
4	5	6	7	8	9	10	11	12	13	14	0	1	2	3
5	6	7	8	9	10	11	12	13	14	0	1	2	3	4
6	7	8	9	10	11	12	13	14	0	1	2	3	4	5
7	8	9	10	11	12	13	14	0	1	2	3	4	5	6
8	9	10	11	12	13	14	0	1	2	3	4	5	6	7
9	10	11	12	13	14	0	1	2	3	4	5	6	7	8
10	11	12	13	14	0	1	2	3	4	5	6	7	8	9
11	12	13	14	0	1	2	3	4	5	6	7	8	9	10
12	13	14	0	1	2	3	4	5	6	7	8	9	10	11
13	14	0	1	2	3	4	5	6	7	8	9	10	11	12
14	0	1	2	3	4	5	6	7	8	9	10	11	12	13

Addition modulo 16

0	1	2	3	4	5	6	7	8	9	10	11	12	13	14	15
1	2	3	4	5	6	7	8	9	10	11	12	13	14	15	0
2	3	4	5	6	7	8	9	10	11	12	13	14	15	0	1
3	4	5	6	7	8	9	10	11	12	13	14	15	0	1	2
4	5	6	7	8	9	10	11	12	13	14	15	0	1	2	3
5	6	7	8	9	10	11	12	13	14	15	0	1	2	3	4
6	7	8	9	10	11	12	13	14	15	0	1	2	3	4	5
7	8	9	10	11	12	13	14	15	0	1	2	3	4	5	6
8	9	10	11	12	13	14	15	0	1	2	3	4	5	6	7
9	10	11	12	13	14	15	0	1	2	3	4	5	6	7	8
10	11	12	13	14	15	0	1	2	3	4	5	6	7	8	9
11	12	13	14	15	0	1	2	3	4	5	6	7	8	9	10
12	13	14	15	0	1	2	3	4	5	6	7	8	9	10	11
13	14	15	0	1	2	3	4	5	6	7	8	9	10	11	12
14	15	0	1	2	3	4	5	6	7	8	9	10	11	12	13
15	0	1	2	3	4	5	6	7	8	9	10	11	12	13	14

28 Does the sequence 1, 1+1, 1+1+1, ... run through all the residue classes, whatever the modulus? Is this inescapable? Can you give a reason?

29 Does the sequence 2, 2+2, 2+2+2, ... run through all the residue classes, whatever the modulus? Is there a pattern which links those moduli for which this sequence completely covers the residue classes, and a pattern which links those moduli for which it does not? Can you prove it?

30 Does the sequence 3, 3+3, 3+3+3, ... run through all the residue classes, whatever the modulus? Is there a pattern which links those moduli for which this sequence completely covers the residue classes, and a pattern which links those moduli for which it does not? Can you prove it?

31 For which moduli $n=7, 8, \ldots, 16$ will the sequence 6, 6+6, 6+6+6, ... reach every residue class?

32 If the sequence a, $a+a$, $a+a+a$, ... reaches every element of $\mathbf{Z}_n$, then a is a *generator* of the group ($\mathbf{Z}_n$, +). List the generators of ($\mathbf{Z}_2$, +), ($\mathbf{Z}_4$, +), ($\mathbf{Z}_8$, +), ($\mathbf{Z}_{16}$, +).
Explain why no even number can generate any of these groups.
Conjecture the number of generators of ($\mathbf{Z}_{2^k}$, +).

33 List the generators of ($\mathbf{Z}_3$, +), ($\mathbf{Z}_9$, +), ($\mathbf{Z}_{27}$, +).
Explain why no multiple of 3 can generate any of these groups.
Conjecture the number of generators of ($\mathbf{Z}_{3^k}$, +).

34 If the sequence a, $a+a$, $a+a+a$, ... is being explored modulo n, can you be sure that the sequence repeats itself after n steps?

35 In the sequences which you wrote down under q 27, what was the first number to be repeated in each sequence?

36 Suppose there is repetition in the sequence

$a, 2a, 3a, \ldots$ modulo n

before reaching $na+a$. Suppose in particular that the first repetition is the term ka which repeats an earlier term la, in the sense that $ka \equiv la \pmod{n}$, and $1 \leq l < k$. Show that a repetition must have occurred at the $(k-l+1)$th term and so $l=1$.

37 If the first repetition in the sequence

$a, 2a, 3a, \ldots$ modulo n

is $ka \equiv a \pmod{n}$, and $1 < k \leq n$, prove that a and n must have a prime factor in common.

⟦42⟧

Table 2.3

1	2	3	4	5	6	7	8
9	10	11	12	13	14	15	16
17	18	19	20	21	22	23	24
25	26	27	28	29	30	31	32
33	34	35	36	37	38	39	40
41	42	43	44	45	46	47	48
49	50	51	52	53	54	55	56
57	58	59	60	61	62	63	64
65	66	67	68	69	70	71	72
73	74	75	76	77	78	79	80
81	82	83	84	85	86	87	88
89	90	91	92	93	94	95	96
97	98	99	100	101	102	103	104
105	106	107	108	109	110	111	112
113	114	115	116	117	118	119	120
121	122	123	124	125	126	127	128

Table 2.4

1	2	3	4	5	6	7	8	9
10	11	12	13	14	15	16	17	18
19	20	21	22	23	24	25	26	27
28	29	30	31	32	33	34	35	36
37	38	39	40	41	42	43	44	45
46	47	48	49	50	51	52	53	54
55	56	57	58	59	60	61	62	63
64	65	66	67	68	69	70	71	72
73	74	75	76	77	78	79	80	81
82	83	84	85	86	87	88	89	90
91	92	93	94	95	96	97	98	99
100	101	102	103	104	105	106	107	108
109	110	111	112	113	114	115	116	117
118	119	120	121	122	123	124	125	126
127	128	129	130	131	132	133	134	135
136	137	138	139	140	141	142	143	144
145	146	147	148	149	150	151	152	153
154	155	156	157	158	159	160	161	162
163	164	165	166	167	168	169	170	171
172	173	174	175	176	177	178	179	180
181	182	183	184	185	186	187	188	189
190	191	192	193	194	195	196	197	198
199	200	201	202	203	204	205	206	207
208	209	210	211	212	213	214	215	216
217	218	219	220	221	222	223	224	225
226	227	228	229	230	231	232	233	234
235	236	237	238	239	240	241	242	243

38 If gcd $(a, n) \neq 1$, prove that a is not a generator of $(\mathbf{Z}_n, +)$.
If gcd $(a, n) = 1$, prove that the sequence $a, 2a, 3a, \ldots$ modulo n does not repeat before the $(n+1)$th term, and deduce that a is a generator of $(\mathbf{Z}_n, +)$.

Euler's ϕ function

39 *Euler's function*, $\phi(n)$, denotes the number of generators of $(\mathbf{Z}_n, +)$, or equivalently, the number of positive integers $\leqslant n$ which are coprime to n. Write down the values of $\phi(n)$ for $n = 1, 2, \ldots, 10$.

40 Use table 2.3 to write down the values of

$\phi(2)$, $\phi(4)$, $\phi(8)$, $\phi(16)$, $\phi(32)$, $\phi(64)$, $\phi(128)$, $\phi(256)$.

Determine the value of $\phi(2^n)$.

41 Use table 2.4 to write down the values of

$\phi(3)$, $\phi(9)$, $\phi(27)$, $\phi(81)$, $\phi(243)$.

Determine the value of $\phi(3^n)$.

42 Use table 2.5 to write down the values of

$\phi(5)$, $\phi(25)$, $\phi(125)$.

Determine the value of $\phi(5^n)$.

Table 2.5

1	2	3	4	5
6	7	8	9	10
11	12	13	14	15
16	17	18	19	20
21	22	23	24	25
26	27	28	29	30
31	32	33	34	35
36	37	38	39	40
41	42	43	44	45
46	47	48	49	50
51	52	53	54	55
56	57	58	59	60
61	62	63	64	65
66	67	68	69	70
71	72	73	74	75
76	77	78	79	80
81	82	83	84	85
86	87	88	89	90
91	92	93	94	95
96	97	98	99	100
101	102	103	104	105
106	107	108	109	110
111	112	113	114	115
116	117	118	119	120
121	122	123	124	125

⟦44⟧

43 State the value of $\phi(p^n)$ for any prime number p and justify your statement.

44 Find $\phi(2)\phi(6)$ and $\phi(3)\phi(4)$. Are either of these products equal to $\phi(12)$?

45 Find $\phi(2)\phi(9)$ and $\phi(3)\phi(6)$. Are either of these two products equal to $\phi(18)$?

46 Prove that, in general, $\phi(ab) \neq \phi(a)\phi(b)$, and make a conjecture about values of a and b for which $\phi(ab) = \phi(a)\phi(b)$.

47 The table found for q 6 was

		Modulo 4 0	1	2	3
	0	0	9	6	3
Modulo 3	1	4	1	10	7
	2	8	5	2	11

Where do the numbers with a factor 3 lie, in this array?
Where do the numbers with a factor 2 lie, in this array?
Must any number which is not coprime to 12 have a factor 2 or 3?
Delete those numbers with a factor 2 or a factor 3 from this array.
How many rows remain which are not wholly deleted?
How many columns remain which are not wholly deleted?
How do these numbers of rows and columns relate to $\phi(12)$?

48 By deleting appropriate rows and columns, use the table given below to illustrate $\phi(36) = \phi(4)\phi(9)$.

		Modulo 9 0	1	2	3	4	5	6	7	8
	0	0	28	20	12	4	32	24	16	8
Modulo 4	1	9	1	29	21	13	5	33	25	17
	2	18	10	2	30	22	14	6	34	26
	3	27	19	11	3	31	23	15	7	35

49 If an integer is *not* coprime to $432 = 16 \cdot 27$, must it have a factor 2 or a factor 3 or both?
Prove that an integer *not* coprime to 432 is congruent to 0, 2, 4, 6, 8, 10, 12, or 14 modulo 16, or else congruent to 0, 3, 6, 9, 12, 15, 18, 21, or 24 modulo 27.
Deduce that $\phi(16 \cdot 27) = \phi(16)\phi(27)$.

50 If m and n are coprime, prove that any number not coprime to mn, either shares a prime factor with m, or shares a prime factor with n, or both. Deduce that if the mn numbers $0, 1, \ldots, mn-1$ are displayed in an $m \times n$ rectangular array with rows congruent modulo m and columns congruent modulo n, and all the numbers in the array not coprime to mn are deleted, then any deleted number lies either in a wholly deleted row, or in a wholly deleted column, or both. Deduce that $\phi(mn) = \phi(m)\phi(n)$.
This is what is meant when ϕ is said to be *multiplicative.*

51 Use q 50 to prove that $\phi(2^a 3^b 5^c) = \phi(2^a)\phi(3^b)\phi(5^c)$.

52 Prove by induction that if $p_1, p_2, \ldots, p_n$ are distinct prime numbers, then

$$\phi(p_1^{\alpha_1} p_2^{\alpha_2} \ldots p_n^{\alpha_n}) = \phi(p_1^{\alpha_1})\phi(p_2^{\alpha_2}) \ldots \phi(p_n^{\alpha_n}).$$

53 Use q 43 to prove that

$$\phi(p_1^{\alpha_1} p_2^{\alpha_2} \ldots p_n^{\alpha_n}) = p_1^{\alpha_1} p_2^{\alpha_2} \ldots p_n^{\alpha_n}\left(1-\frac{1}{p_1}\right)\left(1-\frac{1}{p_2}\right)\ldots\left(1-\frac{1}{p_n}\right).$$

54 Another way to find $\phi(60)$ would be to consider the set of numbers from 1 to 60, and counting elements of this set only, to say

$\phi(60) = 60 -$ (the number of numbers with a factor 2, 3 or 5)
$= 60 -$ (the number of numbers with a factor 2)
$-$ (the number of numbers with a factor 3)
$-$ (the number of numbers with a factor 5)
$+$ (the number of numbers with a factor 2 and 3)
$+$ (the number of numbers with a factor 2 and 5)
$+$ (the number of numbers with a factor 3 and 5)
$-$ (the number of numbers with a factor 2, 3 and 5).

Find the numerical values of the brackets, and verify that $\phi(60) = \phi(4) \cdot \phi(3) \cdot \phi(5)$.

55 Generalise the argument of q 54 to prove that if p, q, and r are distinct primes, then

$$\phi(p^a q^b r^c) = p^a q^b r^c\left(1 - \frac{1}{p} - \frac{1}{q} - \frac{1}{r} + \frac{1}{qr} + \frac{1}{rp} + \frac{1}{pq} - \frac{1}{pqr}\right).$$

Then use q 43 to prove that $\phi(p^a q^b r^c) = \phi(p^a)\phi(q^b)\phi(r^c)$.

56 Extend the argument of q 55, to evaluate Euler's ϕ function for a number with four prime factors.

⟦45⟧

Summing Euler's function over divisors

57 The factors of 6 are 1, 2, 3, and 6. Find
$\phi(1)+\phi(2)+\phi(3)+\phi(6)$.
The factors of 8 are 1, 2, 4, and 8. Find
$\phi(1)+\phi(2)+\phi(4)+\phi(8)$.
The factors of 12 are 1, 2, 3, 4, 6, and 12. Find
$\phi(1)+\phi(2)+\phi(3)+\phi(4)+\phi(6)+\phi(12)$.
Express these results in the form

$$\sum_{d|6}\phi(d)= \quad , \qquad \sum_{d|8}\phi(d)= \qquad \text{and } \sum_{d|12}\phi(d)= \quad .$$

58 List the numbers from 1 to 12 in a row.
Underline the numbers in this list which are coprime to 12. There are $\phi(12)=4$ of them. Rewrite the numbers which you have not underlined below the first row.
Underline the numbers in this second row which are divisible by 2 and, when divided by 2, are coprime to $12/2=6$. There are $\phi(6)=2$ of them. Rewrite the numbers which you have not underlined below the second row.
Underline the numbers in this third row which are divisible by 3 and, when divided by 3, are coprime to $12/3=4$. There are $\phi(4)=2$ of them. Rewrite the numbers which you have not underlined below the third row.
Underline the numbers in this fourth row which are divisible by 4 and, when divided by 4, are coprime to $12/4=3$. There are $\phi(3)=2$ of them. Rewrite the numbers which you have not underlined below the fourth row.
Underline the numbers in this fifth row which, when divided by 6, are coprime to $12/6=2$. There is $\phi(2)=1$ of them. The only number which has not now been underlined is divisible by 12, and there is just one such number.
Try to characterise those numbers which are underlined in each row.

59 If the numbers $1, 2, \ldots, n$ are listed in a row, and the procedure of question 58 is undertaken, in what row will the number a, $1\leqslant a\leqslant n$, be underlined? Assume that the divisors of n in ascending order are $1=d_1, d_2, \ldots, d_k=n$, and that $\gcd(a, n)=d_i$. How many numbers are underlined in the same row as a?

60 Explain why $\sum_{d|n}\phi(d)=\sum_{d|n}\phi(n/d)$.

⟦46⟧

61 If we define $f(n)=\sum_{d|n}\phi(d)$, use q 43 to prove that $f(p^a)=p^a$, for any prime number p.

62 Prove that $f(9)f(8)=f(72)$, using only the definition of f and the fact that $\phi(m)\phi(n)=\phi(mn)$ when m and n are coprime. Deduce that $f(72)=72$.

63 If m and n are coprime, give a reason to justify each step of the following argument.

$$\begin{aligned} f(m)f(n) &= \sum_{d|m}\phi(d)\cdot\sum_{d'|n}\phi(d') \\ &= \sum_{d|m,d'|n}\phi(d)\phi(d') \\ &= \sum_{d|m,d'|n}\phi(dd') \\ &= \sum_{dd'|mn}\phi(dd') \\ &= f(mn). \end{aligned}$$

64 Use q 61 and q 63 to prove that $f(n)=n$ for all positive integers n.

This result provides an essential step in proving the existence of primitive roots to a prime modulus, q 3.61.

Notes and answers

For concurrent reading see bibliography: Davenport (1968), Ore (1948), Weil (1979).

1 The numbers a and b are in the same column.

2 $a \equiv b \pmod 6$ because $a - a = 0 = 6 \cdot 0$.

$a \equiv b \pmod 6 \Rightarrow a - b = 6k \Rightarrow b - a = 6(-k)$

$\Rightarrow b \equiv a \pmod 6$.

$a \equiv b \pmod 6$ and $b \equiv c \pmod 6$

$\Rightarrow a - b = 6k$ and $b - c = 6m$

$\Rightarrow a - c = 6(k + m) \Rightarrow a \equiv c \pmod 6$.

3 If $a = 6q + r$, with $0 \leqslant r < 6$, then $a \equiv r \pmod 6$.

4 Replace 6 by n in q 2.

5

0	4	2
3	1	5

6

0	9	6	3
4	1	10	7
8	5	2	11

7

0	28	20	12	4	32	24	16	8
9	1	29	21	13	5	33	25	17
18	10	2	30	22	14	6	34	26
27	19	11	3	31	23	15	7	35

8 If a and b are in the same position, then $a \equiv b \pmod{16}$ and $a \equiv b \pmod{27}$, so $a - b = 16n$ and $a - b = 27m$. Thus $a - b = 16 \cdot 27 \cdot k$ and $a \equiv b \pmod{432}$. So there is at most one number in each position. But there are 432 numbers and 432 positions, so the matching is one–one.

9

0 6	–	2 8	–	4 10	–
–	1 7	–	3 9	–	5 11

6 and 2 have a common factor, so $a \equiv b \pmod 6$ and $a \equiv b \pmod 2$ only implies $a \equiv b \pmod 6$, not $a \equiv b \pmod{12}$.

10

0 30	–	12 42	–	24 54	–	6 36	–	18 48	–
–	1 31	–	13 43	–	25 55	–	7 37	–	19 49
20 50	–	2 32	–	14 44	–	26 56	–	8 38	–
–	21 51	–	3 33	–	15 45	–	27 57	–	9 39
10 40	–	22 52	–	4 34	–	16 46	–	28 58	–
–	11 41	–	23 53	–	5 35	–	17 47	–	29 59

10 and 6 have a common factor, so $a \equiv b \pmod{10}$ and $a \equiv b \pmod 6$ only implies $a \equiv b \pmod{30}$, not $a \equiv b \pmod{60}$.

11 The numbers m and n should be coprime.

12 $$\left.\begin{array}{l} a \equiv b \pmod{m} \Rightarrow a-b=mk \\ a \equiv b \pmod{n} \Rightarrow a-b=nh \end{array}\right\} \begin{array}{l} \Rightarrow n|mk \text{ and so } n|k \text{ by n 1.61} \\ \Rightarrow a-b=mnl \\ \Rightarrow a \equiv b \pmod{mn}. \end{array}$$

13 There cannot be two solutions by q 12, so exactly one of the mn numbers $0, 1, \ldots, mn-1$ matches each of the mn choices of (a_1, a_2).

14 14

15 29

16 29

17 By q 13, there is a unique solution $x=b$, $0 \leqslant b \leqslant m_1m_2-1$, to

$x \equiv a_1 \pmod{m_1}$,

$x \equiv a_2 \pmod{m_2}$.

Now we can use q 1.60 and q 13 again to show that there is a unique solution to

$x \equiv b \pmod{m_1m_2}$,

$x \equiv a_3 \pmod{m_3}$,

with $0 \leqslant x \leqslant m_1m_2m_3-1$.

18 Suppose there is a unique solution for $n-1$ such equations, that is a unique $x=b$, $0 \leqslant b \leqslant m_1m_2 \ldots m_{n-1}-1$ satisfying

$x \equiv a_1 \pmod{m_1}$,

$x \equiv a_2 \pmod{m_2}$,

$\vdots$

$x \equiv a_{n-1} \pmod{m_{n-1}}$,

then by q 1.60 and q 13 there is a unique x, $0 \leqslant x \leqslant m_1m_2 \ldots m_n-1$ satisfying

$x \equiv b \pmod{m_1m_2 \ldots m_{n-1}}$

and

$x \equiv a_n \pmod{m_n}$.

The result now follows by induction.

19 Yes.

20 $-12 \equiv 0 \pmod 6$, $-5 \equiv 1 \pmod 6$, $8 \equiv 2 \pmod 6$, $-3 \equiv 3 \pmod 6$, $-8 \equiv 4 \pmod 6$, $23 \equiv 5 \pmod 6$.
$\{6n+8 \mid n \in \mathbf{Z}\} = \{6n+2 \mid n \in \mathbf{Z}\}$ etc.

21 Each is congruent to one of 0, 1, 2, 3, 4, 5. No two may be congruent to the same number, or they will be congruent to each other by the result of q 2. Thus exactly one comes from each residue class.

22 $a \equiv b \pmod 6 \Rightarrow a - b = 6k$
$c \equiv d \pmod 6 \Rightarrow c - d = 6m$

$$(a+c)-(b+d)=6(k+m) \Rightarrow a+c \equiv b+d \pmod 6$$
$$(a-c)-(b-d)=6(k-m) \Rightarrow a-c \equiv b-d \pmod 6$$

23

0	1	2	3	4	5
1	2	3	4	5	0
2	3	4	5	0	1
3	4	5	0	1	2
4	5	0	1	2	3
5	0	1	2	3	4

0 is the identity.
1 generates the group, as does 5.

24

12	−5	8	−3	−8	23
−5	8	−3	−8	23	12
8	−3	−8	23	12	−5
−3	−8	23	12	−5	8
−8	23	12	−5	8	−3
23	12	−5	8	−3	−8

12 is the identity.
−5 generates the group, as does 23.

25 $a_1, \ldots, a_6$ are congruent modulo 6 to 0, 1, 2, 3, 4, 5 in some order. Suppose $a_i \equiv i \pmod 6$, then $a_i + a_j \equiv i + j \pmod 6$, so this table has the same structure as that of q 23. The identity is the $a_i \equiv 0 \pmod 6$ and a generator is the $a_i \equiv 1 \pmod 6$.
Each of the groups obtained in this way is an example of the group $(\mathbf{Z}_6, +)$ which may either be defined on a complete set of residues with addition modulo 6, or may be defined to have the residue classes modulo 6 as its elements and its addition the one induced by the usual addition on the integers on the classes as established in q 22. The two descriptions provide the same algebraic structure.

26 As q 22 with n for 6.

27 See q 28, q 29, q 30.

28

Mod 3	1	2	0	1	2	0	1	2	0	1
Mod 4	1	2	3	0	1	2	3	0	1	2
Mod 5	1	2	3	4	0	1	2	3	4	0
Mod 6	1	2	3	4	5	0	1	2	3	4
Mod 7	1	2	3	4	5	6	0	1	2	3
Mod 8	1	2	3	4	5	6	7	0	1	2
Mod 9	1	2	3	4	5	6	7	8	0	1
Mod 10	1	2	3	4	5	6	7	8	9	0

Yes, every positive integer has the form $1+1+\ldots+1$.

29

Mod 3	2	1	0	2	1	0	2	1	0	2
Mod 4	2	0	2	0	2	0	2	0	2	0
Mod 5	2	4	1	3	0	2	4	1	3	0
Mod 6	2	4	0	2	4	0	2	4	0	2
Mod 7	2	4	6	1	3	5	0	2	4	6
Mod 8	2	4	6	0	2	4	6	0	2	4
Mod 9	2	4	6	8	1	3	5	7	0	2
Mod 10	2	4	6	8	0	2	4	6	8	0

For odd moduli, the sequence 2, 2+2, ... covers all the residue classes.
Modulo $2k$ we have $2, 4, \ldots, 2(k-1), 0, 2, 4, \ldots$
Modulo $2k+1$ we have $2, 4, \ldots, 2k, 1, 3, \ldots, 2k-1, 0,$.

30

Mod 3	0	0	0	0	0	0	0	0	0	0
Mod 4	3	2	1	0	3	2	1	0	3	2
Mod 5	3	1	4	2	0	3	1	4	2	0
Mod 6	3	0	3	0	3	0	3	0	3	0
Mod 7	3	6	2	5	1	4	0	3	6	2
Mod 8	3	6	1	4	7	2	5	0	3	6
Mod 9	3	6	0	3	6	0	3	6	0	3
Mod 10	3	6	9	2	5	8	1	4	7	0

For moduli without a factor 3, the sequence 3, 3+3, ... covers all the residue classes.
Modulo $3k$ we have $3, 6, \ldots, 3(k-1), 0, 3, 6, \ldots$
Modulo $3k+1$ we have $3, 6, \ldots, 3k, 2, 5, \ldots, 3k-1, 1, 4, \ldots, 3k-2, 0, \ldots$
Modulo $3k+2$ we have $3, 6, \ldots, 3k, 1, 4, \ldots, 3k+1, 2, 5, \ldots, 3k-1, 0, 3, \ldots$

31 7, 11, 13.

32 $(\mathbf{Z}_2, +)$ 1
$(\mathbf{Z}_4, +)$ 1, 3
$(\mathbf{Z}_8, +)$ 1, 3, 5, 7
$(\mathbf{Z}_{16}, +)$ 1, 3, 5, 7, 9, 11, 13, 15
No even number, modulo 2^k, can generate an odd number.
The number of generators of $(\mathbf{Z}_{2^k}, +)$ is 2^{k-1}.

33 $(\mathbf{Z}_3, +)$ 1, 2
$(\mathbf{Z}_9, +)$ 1, 2, 4, 5, 7, 8
$(\mathbf{Z}_{27}, +)$ 1, 2, 4, 5, 7, 8, 10, 11, 13, 14, 16, 17, 19, 20, 22, 23, 25, 26
No multiple of 3, modulo 3^k, can generate a number of the form $3n+1$.
The number of generators of $(\mathbf{Z}_{3^k}, +)$ is $2 \cdot 3^{k-1}$.

34 $na + a \equiv a \pmod{n}$, $na + ka \equiv ka \pmod{n}$.

35 The first number of the sequence.

36 If $ka \equiv la \pmod{n}$, then $ka - la \equiv 0 \pmod{n}$, so $ka - la + a \equiv a \pmod{n}$, and $(k-l+1)a \equiv a \pmod{n}$. Since the first repetition is the kth term, $k-l+1 \geqslant k$, and $1 \geqslant l$, so $l=1$.

37 If $ka \equiv a \pmod{n}$, $(k-1)a \equiv 0 \pmod{n}$, so $n | a(k-1)$. If $\gcd(a, n) = 1$, by q 1.61, $n | k-1$. But $0 < k-1 < n$, so n is not a factor of $k-1$. Thus n and a have a common factor.

38 If $\gcd(a, n) = d \neq 1$, then a cannot generate (modulo n) a number which is not a multiple of d.
If $\gcd(a, n) = 1$, then from q 1.61, $n | k-1$, so $k-1 \geqslant n$, $k \geqslant n+1$, so the first n terms of the sequence are distinct, and a generates the group.

The last nineteen questions have been concerned with defining the groups $(\mathbf{Z}_n, +)$ and identifying the generators of these groups. If $\mathbf{0}, \mathbf{1}, \ldots, \boldsymbol{n-1}$ represent the n residue classes modulo n with $a \in \boldsymbol{a}$, then $\boldsymbol{a}$ is a generator of $(\mathbf{Z}_n, +)$ precisely when $\gcd(a, n) = 1$.

39 $\phi(1) = 1$ 1
$\phi(2) = 1$ 1
$\phi(3) = 2$ 1, 2
$\phi(4) = 2$ 1, 3
$\phi(5) = 4$ 1, 2, 3, 4
$\phi(6) = 2$ 1, 5
$\phi(7) = 6$ 1, 2, 3, 4, 5, 6
$\phi(8) = 4$ 1, 3, 5, 7
$\phi(9) = 6$ 1, 2, 4, 5, 7, 8
$\phi(10) = 4$ 1, 3, 7, 9

40 $\phi(2) = 1$, $\phi(4) = 2$, $\phi(8) = 4$, $\phi(16) = 8$, $\phi(32) = 16$, $\phi(64) = 32$, $\phi(128) = 64$, $\phi(256) = 128$. $\phi(2^n) = 2^{n-1}$. Odd numbers generate, even numbers do not.

41 $\phi(3) = 2$, $\phi(9) = 6$, $\phi(27) = 18$, $\phi(81) = 54$, $\phi(243) = 162$. $\phi(3^n) = 2 \cdot 3^{n-1}$. Every third number has a factor 3.

42 $\phi(5) = 4$, $\phi(25) = 20$, $\phi(125) = 100$. $\phi(5^n) = 4 \cdot 5^{n-1}$. Every fifth number has a factor 5.

43 $\phi(p^n) = (p-1)p^{n-1}$. Only multiples of p have a factor in common with p^n, when p is a prime. There are just p^{n-1} of these $\leqslant p^n$, so $\phi(p^n) = p^n - p^{n-1}$.
It is important to memorise this result.

44 $\phi(2)\phi(6) = 2$, $\phi(3)\phi(4) = 2 \cdot 2 = 4 = \phi(12)$.

45 $\phi(2)\phi(9) = 6 = \phi(18)$, $\phi(3)\phi(6) = 4$.

46 $\phi(2)\phi(6) \neq \phi(12)$.
If $\gcd(a, b) = 1$, then $\phi(ab) = \phi(a)\phi(b)$. The proof follows in n 50.

47

(0)	(9)	(6)	(3)
(4)	1	(10)	7
(8)	6	(2)	11

$\phi(3)$ rows remain, $\phi(4)$ columns remain.

48

(0)	(28)	(20)	(12)	(4)	(32)	(24)	(16)	(8)
(9)	1	29	(21)	13	5	(33)	25	17
(18)	(10)	(2)	(30)	(22)	(14)	(6)	(34)	(26)
(27)	19	11	(3)	31	23	(15)	7	35

$\phi(4)$ rows remain, $\phi(9)$ columns remain.

49 If two numbers have a common factor, they must have a prime common factor. Numbers with a factor 2 are just those which are congruent to 0, 2, 4, 6, 8, 10, 12, 14 (mod 16). Numbers with a factor 3 are just those congruent to 0, 3, 6, 9, 12, 15, 18, 21, 24 (mod 27). Any number not coprime to 432 has a factor 2 or a factor 3, or both. The numbers which are coprime to 432 are just those which are congruent to 1, 3, 5, 7, 9, 11, 13, 15 (mod 16) and simultaneously to 1, 2, 4, 5, 7, 8, 10, 11, 13, 14, 16, 17, 19, 20, 22, 23, 25, 26 (mod 27).
So there are $\phi(16)\phi(27)$ of them.

50 If $\gcd(a, mn) = d \neq 1$, then d has a prime factor p (say) and $p|mn$ so $p|m$ or $p|n$.
If $p|m$, then each number in the row containing a has a factor p, for $a \equiv b \pmod m \Rightarrow b = a + mk$, so $p|b$.
If $p|n$, then each number in the column containing a has a factor p, similarly.
Number of rows not wholly deleted $= \phi(m)$.
Number of columns not wholly deleted $= \phi(n)$.
So $\phi(mn) = \phi(m)\phi(n)$.

Once we know that $\phi(p^n) = p^n - p^{n-1}$ for any prime p, and that ϕ is multiplicative, we can use the fundamental theorem of arithmetic to determine the value of $\phi(n)$ for any n. This technique is worth remembering.

51 $\phi(2^a 3^b 5^c) = \phi(2^a 3^b)\phi(5^c) = \phi(2^a)\phi(3^b)\phi(5^c)$.

52 Suppose $\phi(p_1^{\alpha_1} \ldots p_{n-1}^{\alpha_{n-1}}) = \phi(p_1^{\alpha_1}) \ldots \phi(p_{n-1}^{\alpha_{n-1}})$. The extension to the nth factor follows by q 50, so the complete result follows by induction.

54 60/2 numbers with a factor 2,
60/3 numbers with a factor 3,
60/5 numbers with a factor 5,
60/(2 · 3) numbers with a factor 2 and a factor 3,
60/(2 · 5) numbers with a factor 2 and a factor 5,
60/(3 · 5) numbers with a factor 3 and a factor 5,
60/(2 · 3 · 5) numbers with factors 2, 3 and 5.

$$\begin{aligned}\phi(60) &= 60-60/2-60/3-60/5+60/3\cdot 5\\ &\quad +60/2\cdot 5+60/2\cdot 3-60/2\cdot 3\cdot 5\\ &= 60(1-\tfrac{1}{2})(1-\tfrac{1}{3})(1-\tfrac{1}{5})\\ &= 4(1-\tfrac{1}{2})3(1-\tfrac{1}{3})5(1-\tfrac{1}{5})\\ &= \phi(4)\phi(3)\phi(5).\end{aligned}$$

56 $$\begin{aligned}\phi(p^a q^b r^c s^d) &= p^a q^b r^c s^d\left(1-\frac{1}{p}-\frac{1}{q}-\frac{1}{r}-\frac{1}{s}+\frac{1}{pq}+\frac{1}{pr}+\frac{1}{ps}+\frac{1}{qr}\right.\\ &\qquad \left.+\frac{1}{qs}+\frac{1}{rs}-\frac{1}{qrs}-\frac{1}{rsp}-\frac{1}{spq}-\frac{1}{pqr}+\frac{1}{pqrs}\right)\\ &= p^a q^b r^c s^d\left(1-\frac{1}{p}\right)\left(1-\frac{1}{q}\right)\left(1-\frac{1}{r}\right)\left(1-\frac{1}{s}\right).\end{aligned}$$

This question completes our investigation of the value of $\phi(n)$.

57 $\sum_{d|6}\phi(d)=6, \quad \sum_{d|8}\phi(d)=8, \quad \sum_{d|12}\phi(d)=12.$

58

$\underline{1}$	2	3	4	$\underline{5}$	6	$\underline{7}$	8	9	10	$\underline{11}$	12
	$\underline{2}$	3	4		6		8	9	$\underline{10}$		12
		$\underline{3}$	4		6		8	$\underline{9}$			12
			$\underline{4}$		6		$\underline{8}$				12
					$\underline{6}$						12
											$\underline{12}$

If gcd $(a, 12)=d$ and d is the ith divisor of 12 in increasing order, then a is underlined in the ith row.

A graphical illustration of this exercise is given in fig. 2.1. Lines of slope $n/12$ have been drawn through the origin for $n=1, 2, \ldots, 12$. On each of these lines, the lattice point nearest to the origin has been encircled.

59 The number a is underlined in the ith row if gcd $(a, n)=d_i$. The set of numbers $\{x \mid 1\leq x\leq n, \text{gcd}(x, n)=d_i\}$ is in one–one correspondence with the numbers coprime to n/d_i of which there are $\phi(n/d_i)$.

60 $\sum_{d|12}\phi(d)=\phi(1)+\phi(2)+\phi(3)+\phi(4)+\phi(6)+\phi(12)$

$\sum_{d|12}\phi(12/d)=\phi(12)+\phi(6)+\phi(4)+\phi(3)+\phi(2)+\phi(1).$

61 $$\begin{aligned}f(p^\alpha) &= \phi(p^\alpha)+\phi(p^{\alpha-1})+\ldots+\phi(p)+\phi(1)\\ &= (p^\alpha-p^{\alpha-1})+(p^{\alpha-1}-p^{\alpha-2})+\ldots+(p-1)+1\\ &= p^\alpha.\end{aligned}$$

Having established that $f(p^\alpha)=p^\alpha$, we can use the fundamental theorem of arithmetic to determine $f(n)$ if only we can prove that f is multiplicative.

Fig. 2.1

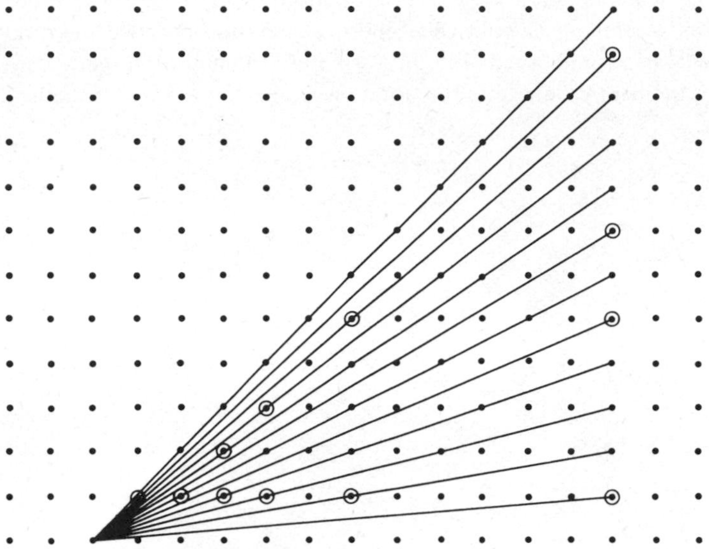

62 $f(9)f(8) = (\phi(1)+\phi(3)+\phi(9))(\phi(1)+\phi(2)+\phi(4)+\phi(8))$

$$\begin{aligned} &= \phi(1)\phi(1)+\phi(1)\phi(2)+\phi(1)\phi(4)+\phi(1)\phi(8) \\ &\quad +\phi(3)\phi(1)+\phi(3)\phi(2)+\phi(3)\phi(4)+\phi(3)\phi(8) \\ &\quad +\phi(9)\phi(1)+\phi(9)\phi(2)+\phi(9)\phi(4)+\phi(9)\phi(8) \\ &= \phi(1)+\phi(2)+\phi(4)+\phi(8) \\ &\quad +\phi(3)+\phi(6)+\phi(12)+\phi(24) \\ &\quad +\phi(9)+\phi(18)+\phi(36)+\phi(72) \\ &= f(72). \end{aligned}$$

63 (i) definition, (ii) algebraic manipulation, (iii) ϕ multiplicative; m, n coprime, (iv) m, n coprime, (v) definition.

The result here establishes that f is multiplicative.

64 $f(p_1^{\alpha_1}\dots p_n^{\alpha_n}) = f(p_1^{\alpha_1})\dots f(p_n^{\alpha_n})$ by q 63

$= p_1^{\alpha_1}\dots p_n^{\alpha_n}$ by q 61.

Historical note

The Chinese remainder theorem is a general solution of a form of problem posed in China and elsewhere about the beginning of the Christian era.

While L. Euler published the first arithmetical discoveries using $\phi(n)$ in 1760, the actual notation '$\phi(n)$' was introduced by C. F. Gauss in his *Disquisitiones Arithmeticae* in 1801 and it was in this book that the result $\sum_{d|n} \phi(d) = n$ was published. The discussion of arithmetical properties in terms of quotient groups, and the notation $\mathbf{Z}_n$, belong to the twentieth century.

3

Modular multiplication

Fermat's theorem

1 $6 \equiv 14 \pmod 4$ and $3 \equiv -1 \pmod 4$. Is it true that $6 \cdot 3 \equiv 14(-1) \pmod 4$?
If $a \equiv b \pmod 4$ and $c \equiv d \pmod 4$, can you prove that $ac \equiv bd \pmod 4$?

2 Make a multiplication table modulo 4 for the complete set of residues $\{-4, 5, 2, -1\}$. Does it have the same structure as the table you found for q 1.14?

3 If $a \equiv b \pmod n$ and $c \equiv d \pmod n$, use the equality $ac - bd = (a-b)c + b(c-d)$ to prove that $ac \equiv bd \pmod n$.

4 If $a_1, \ldots, a_n$ is a complete set of residues modulo n, and the product modulo n of a_i and a_j is defined to be a_k where $a_i a_j \equiv a_k \pmod n$, is a_k uniquely defined?
Is there an element a_Z such that $a_Z a_i \equiv a_Z$ for all i?
Is there an element a_I such that $a_I a_i \equiv a_i$ for all i?

5 If $a_1, \ldots, a_n$ is a complete set of residues modulo n, and $b_1, \ldots, b_n$ is a complete set of residues modulo n, how would you establish that the tables of products modulo n obtained from these two sets have the same structure? Your answer establishes a well-defined table for $(\mathbf{Z}_n, \times)$.

6 Examine the tables for multiplication modulo 3, 4, ..., 16 given in table 3.1. Is there an identity in $(\mathbf{Z}_n, \times)$, for every n?

7 Compare the second row of the tables for $(\mathbf{Z}_3, \times)$, $(\mathbf{Z}_4, \times), \ldots, (\mathbf{Z}_{10}, \times)$ with your answers to q 2.28. What is the similarity?

8 Compare the third row of the tables for $(\mathbf{Z}_3, \times), \ldots, (\mathbf{Z}_{10}, \times)$ with your answers to q 2.29. What is the similarity?

9 Compare the fourth row of the tables for $(\mathbf{Z}_3, \times), \ldots, (\mathbf{Z}_{10}, \times)$ with your answers to q 2.30. What is the similarity?

Table 3.1

Multiplication modulo 3

0	0	0
0	1	2
0	2	1

Multiplication modulo 4

0	0	0	0
0	1	2	3
0	2	0	2
0	3	2	1

Multiplication modulo 5

0	0	0	0	0
0	1	2	3	4
0	2	4	1	3
0	3	1	4	2
0	4	3	2	1

Multiplication modulo 6

0	0	0	0	0	0
0	1	2	3	4	5
0	2	4	0	2	4
0	3	0	3	0	3
0	4	2	0	4	2
0	5	4	3	2	1

Multiplication modulo 7

0	0	0	0	0	0	0
0	1	2	3	4	5	6
0	2	4	6	1	3	5
0	3	6	2	5	1	4
0	4	1	5	2	6	3
0	5	3	1	6	4	2
0	6	5	4	3	2	1

Multiplication modulo 8

0	0	0	0	0	0	0	0
0	1	2	3	4	5	6	7
0	2	4	6	0	2	4	6
0	3	6	1	4	7	2	5
0	4	0	4	0	4	0	4
0	5	2	7	4	1	6	3
0	6	4	2	0	6	4	2
0	7	6	5	4	3	2	1

⟦69⟧

Table 3.1 (*cont.*)

Multiplication modulo 9

0	0	0	0	0	0	0	0	0
0	1	2	3	4	5	6	7	8
0	2	4	6	8	1	3	5	7
0	3	6	0	3	6	0	3	6
0	4	8	3	7	2	6	1	5
0	5	1	6	2	7	3	8	4
0	6	3	0	6	3	0	6	3
0	7	5	3	1	8	6	4	2
0	8	7	6	5	4	3	2	1

Multiplication modulo 10

0	0	0	0	0	0	0	0	0	0
0	1	2	3	4	5	6	7	8	9
0	2	4	6	8	0	2	4	6	8
0	3	6	9	2	5	8	1	4	7
0	4	8	2	6	0	4	8	2	6
0	5	0	5	0	5	0	5	0	5
0	6	2	8	4	0	6	2	8	4
0	7	4	1	8	5	2	9	6	3
0	8	6	4	2	0	8	6	4	2
0	9	8	7	6	5	4	3	2	1

Multiplication modulo 11

0	0	0	0	0	0	0	0	0	0	0
0	1	2	3	4	5	6	7	8	9	10
0	2	4	6	8	10	1	3	5	7	9
0	3	6	9	1	4	7	10	2	5	8
0	4	8	1	5	9	2	6	10	3	7
0	5	10	4	9	3	8	2	7	1	6
0	6	1	7	2	8	3	9	4	10	5
0	7	3	10	6	2	9	5	1	8	4
0	8	5	2	10	7	4	1	9	6	3
0	9	7	5	3	1	10	8	6	4	2
0	10	9	8	7	6	5	4	3	2	1

Multiplication modulo 12

0	0	0	0	0	0	0	0	0	0	0	0
0	1	2	3	4	5	6	7	8	9	10	11
0	2	4	6	8	10	0	2	4	6	8	10
0	3	6	9	0	3	6	9	0	3	6	9
0	4	8	0	4	8	0	4	8	0	4	8
0	5	10	3	8	1	6	11	4	9	2	7
0	6	0	6	0	6	0	6	0	6	0	6
0	7	2	9	4	11	6	1	8	3	10	5
0	8	4	0	8	4	0	8	4	0	8	4
0	9	6	3	0	9	6	3	0	9	6	3
0	10	8	6	4	2	0	10	8	6	4	2
0	11	10	9	8	7	6	5	4	3	2	1

Table 3.1 (*cont.*)

Multiplication modulo 13

0	0	0	0	0	0	0	0	0	0	0	0	0
0	1	2	3	4	5	6	7	8	9	10	11	12
0	2	4	6	8	10	12	1	3	5	7	9	11
0	3	6	9	12	2	5	8	11	1	4	7	10
0	4	8	12	3	7	11	2	6	10	1	5	9
0	5	10	2	7	12	4	9	1	6	11	3	8
0	6	12	5	11	4	10	3	9	2	8	1	7
0	7	1	8	2	9	3	10	4	11	5	12	6
0	8	3	11	6	1	9	4	12	7	2	10	5
0	9	5	1	10	6	2	11	7	3	12	8	4
0	10	7	4	1	11	8	5	2	12	9	6	3
0	11	9	7	5	3	1	12	10	8	6	4	2
0	12	11	10	9	8	7	6	5	4	3	2	1

Multiplication modulo 14

0	0	0	0	0	0	0	0	0	0	0	0	0	0
0	1	2	3	4	5	6	7	8	9	10	11	12	13
0	2	4	6	8	10	12	0	2	4	6	8	10	12
0	3	6	9	12	1	4	7	10	13	2	5	8	11
0	4	8	12	2	6	10	0	4	8	12	2	6	10
0	5	10	1	6	11	2	7	12	3	8	13	4	9
0	6	12	4	10	2	8	0	6	12	4	10	2	8
0	7	0	7	0	7	0	7	0	7	0	7	0	7
0	8	2	10	4	12	6	0	8	2	10	4	12	6
0	9	4	13	8	3	12	7	2	11	6	1	10	5
0	10	6	2	12	8	4	0	10	6	2	12	8	4
0	11	8	5	2	13	10	7	4	1	12	9	6	3
0	12	10	8	6	4	2	0	12	10	8	6	4	2
0	13	12	11	10	9	8	7	6	5	4	3	2	1

Multiplication modulo 15

0	0	0	0	0	0	0	0	0	0	0	0	0	0	0
0	1	2	3	4	5	6	7	8	9	10	11	12	13	14
0	2	4	6	8	10	12	14	1	3	5	7	9	11	13
0	3	6	9	12	0	3	6	9	12	0	3	6	9	12
0	4	8	12	1	5	9	13	2	6	10	14	3	7	11
0	5	10	0	5	10	0	5	10	0	5	10	0	5	10
0	6	12	3	9	0	6	12	3	9	0	6	12	3	9
0	7	14	6	13	5	12	4	11	3	10	2	9	1	8
0	8	1	9	2	10	3	11	4	12	5	13	6	14	7
0	9	3	12	6	0	9	3	12	6	0	9	3	12	6
0	10	5	0	10	5	0	10	5	0	10	5	0	10	5
0	11	7	3	14	10	6	2	13	9	5	1	12	8	4
0	12	9	6	3	0	12	9	6	3	0	12	9	6	3
0	13	11	9	7	5	3	1	14	12	10	8	6	4	2
0	14	13	12	11	10	9	8	7	6	5	4	3	2	1

Table 3.1 (*cont.*)

Multiplication modulo 16

0	0	0	0	0	0	0	0	0	0	0	0	0	0	0	0
0	1	2	3	4	5	6	7	8	9	10	11	12	13	14	15
0	2	4	6	8	10	12	14	0	2	4	6	8	10	12	14
0	3	6	9	12	15	2	5	8	11	14	1	4	7	10	13
0	4	8	12	0	4	8	12	0	4	8	12	0	4	8	12
0	5	10	15	4	9	14	3	8	13	2	7	12	1	6	11
0	6	12	2	8	14	4	10	0	6	12	2	8	14	4	10
0	7	14	5	12	3	10	1	8	15	6	13	4	11	2	9
0	8	0	8	0	8	0	8	0	8	0	8	0	8	0	8
0	9	2	11	4	13	6	15	8	1	10	3	12	5	14	7
0	10	4	14	8	2	12	6	0	10	4	14	8	2	12	6
0	11	6	1	12	7	2	13	8	3	14	9	4	15	10	5
0	12	8	4	0	12	8	4	0	12	8	4	0	12	8	4
0	13	10	7	4	1	14	11	8	5	2	15	12	9	6	3
0	14	12	10	8	6	4	2	0	14	12	10	8	6	4	2
0	15	14	13	12	11	10	9	8	7	6	5	4	3	2	1

10 What is the similarity between the $(a+1)$th row of the table for $(\mathbf{Z}_n, \times)$, and the terms of the sequence $a, a+a, a+a+a, \ldots \pmod{n}$?

11 What is the term which precedes the first repetition in the sequence $a, a+a, a+a+a, \ldots \pmod{n}$?

12 In table 3.1, for which moduli $n = 3, \ldots, 16$ do zeros appear in $(\mathbf{Z}_n, \times)$ other than in the first row or first column?

13 If n is a composite number, give a reason why there must be a zero in $(\mathbf{Z}_n, \times)$ other than in the first row or column. Explain why a row or column containing two zeros cannot contain all the elements of $\mathbf{Z}_n$.

14 If p is a prime number, use q 2.38 to explain why each row of the table for $(\mathbf{Z}_p, \times)$ contains all the elements of $\mathbf{Z}_p$. This is as much as to say that if $0 < a < p$, then $x \to ax$ is a permutation of the elements of $\mathbf{Z}_p$.
Deduce that $ax \equiv ay \pmod{p}$ implies $x \equiv y \pmod{p}$, and that if, in addition $0 < b < p$, there is a unique x between 0 and p such that $ax \equiv b \pmod{p}$.

15 If p is a prime number and $0 < a < p$ do each of the sets $\{1, 2, \ldots, p-1\}$ and $\{a, 2a, \ldots, (p-1)a\}$ include all the non-zero elements of $\mathbf{Z}_p$?
What can you deduce about the product of all the elements in one set compared with the product of all the elements in the

other? Use the cancellation property established in q 14 to show that $a^{p-1} \equiv 1 \pmod p$.

16 Deduce from q 14 that in $(\mathbf{Z}_p, \times)$ every element except 0 has a multiplicative inverse. That is, for each $x \not\equiv 0 \pmod p$, there exists an $a \not\equiv 0 \pmod p$ such that $ax \equiv 1 \pmod p$.

17 Do the non-zero elements of $\mathbf{Z}_p$ form a group under multiplication mod p?

18 Use the theorem of Lagrange, that the order of a subgroup divides the order of a group, to prove that for all non-zero elements of $\mathbf{Z}_p$, $x^{p-1} = 1$.

19 From q 15 or q 18, deduce that $x^p \equiv x \pmod p$ for all integers x and all prime numbers p. (*Fermat's theorem*)

20 Prove that $2222^{5555} + 5555^{2222}$ is divisible by 7. Invent a similar problem.

Wilson's theorem

21 For a prime number p, we denote the set of non-zero elements of $\mathbf{Z}_p$ by $\mathbf{M}_p$, and then $(\mathbf{M}_p, \times)$ forms a group. In the groups $\mathbf{M}_3$, $\mathbf{M}_5$, $\mathbf{M}_7$, $\mathbf{M}_{11}$ and $\mathbf{M}_{13}$, find the elements x such that $x^2 = 1$. Generalise your result and identify the elements x such that $x^2 = 1$ in $\mathbf{M}_p$.

22 Use the fact that $2 \cdot 4 = 1$ and $3 \cdot 5 = 1$ in $\mathbf{M}_7$ to prove that $1 \cdot 2 \cdot 3 \cdot 4 \cdot 5 \cdot 6 \equiv 6 \pmod 7$.

23 Evaluate $1 \cdot 2 \cdot 3 \cdot 4 \cdot 5 \cdot 6 \cdot 7 \cdot 8 \cdot 9 \cdot 10 \pmod{11}$.

24 How many of the elements $1, 2, 3, \ldots, p-1$ can be paired with a distinct multiplicative inverse in $\mathbf{M}_p$?

25 Evaluate the product $(p-1)! \pmod p$. (*Wilson's theorem*)

Linear congruences

26 Use table 3.1 to determine the solutions, if any, to

$4x \equiv 10 \pmod{14}$,

$4x \equiv 9 \pmod{14}$,

$9x \equiv 2 \pmod{14}$.

27 For what values of b does

$4x \equiv b \pmod{14}$

have (i) more than one solution, with $0 \leqslant x < 14$, (ii) no solutions?

⟦70⟧

28 If b is an odd number, give a formal proof that there are no integers x such that $4x \equiv b \pmod{14}$.

29 If b is an even number, $b = 2c$, say, is the congruence $4x \equiv 2c \pmod{14}$ equivalent to the congruence $2x \equiv c \pmod 7$? Show that the unique solution mod 7 to $2x \equiv c \pmod 7$ gives rise to two solutions of $4x \equiv b \pmod{14}$.

30 If $ax \equiv b \pmod n$ has a solution for x, prove that $\gcd(a, n)|b$.

31 If $\gcd(a, n) = 1$, then from q 1.34 there are integers r and s such that $ar + ns = 1$. Deduce that $ax \equiv b \pmod n$ has a solution br, whatever the value of b.

32 If $\gcd(a, n) = 1$, $ax \equiv b \pmod n$ and $ay \equiv b \pmod n$, prove that $x \equiv y \pmod n$.

33 Use q 31 and q 32 to prove that if $\gcd(a, n) = d|b$, then $(a/d)x \equiv b/d \pmod{n/d}$ has a unique solution modulo n/d. Deduce that $ax \equiv b \pmod n$ has d solutions which are not congruent modulo n.

Fermat–Euler theorem

We now devise a generalisation of Fermat's theorem which holds whether the modulus is prime or not. This depends upon finding a group within $(\mathbf{Z}_n, \times)$.

34 Use table 3.1 to list the values of a for which

$ax \equiv 1 \quad \pmod{14}$

has a solution.
Use table 3.1 to list the values of a for which

$x \to ax$

is a permutation of the elements of $\mathbf{Z}_{14}$.
Can you give a reason for the similarity between these two sets?

35 Use table 3.1 to list the values of a for which

$ax \equiv 1 \quad \pmod{12}$

has a solution.
Use table 3.1 to list the values of a for which

$x \to ax$

is a permutation of the elements of $\mathbf{Z}_{12}$.

36 Establish the equivalence of the following three subsets of $\{1, 2, \ldots, n-1\}$ for $n \geq 2$:

[[71]]

(i) $\{a \mid \gcd(a, n) = 1\}$,
(ii) $\{a \mid ax \equiv 1 \pmod{n}$ has a solution$\}$,
(iii) $\{a \mid x \to ax$ is a permutation of $\mathbf{Z}_n\}$.

How many elements are there in each of these sets? Such a set is said to form a *reduced* set of residues modulo n, and we shall denote it by $\mathbf{M}_n$.

37 Verify that when n is a prime number, the definition of $\mathbf{M}_n$ in q 36 accords with the definition of $\mathbf{M}_p$ in q 21.

38 Give the multiplication tables for the sets $\mathbf{M}_n$ when $n = 4, 6, 8, 9, 10$.

39 Is each of the tables you have found in q 38, that of a group?

40 Use the second definition of $\mathbf{M}_n$ in q 36 to prove that $(\mathbf{M}_n, \times)$ is a group.

41 If $a_1, a_2, \ldots, a_{\phi(n)}$ is a reduced set of residues modulo n, use q 40 to prove that $a_ia_1, a_ia_2, \ldots, a_ia_{\phi(n)}$ is also a reduced set of residues modulo n, and by considering the products of these sets prove that $a_i^{\phi(n)} \equiv 1 \pmod{n}$.
(*Fermat–Euler theorem*)

42 Use Lagrange's theorem on subgroups to prove that if $\gcd(a, n) = 1$, then $a^{\phi(n)} \equiv 1 \pmod{n}$.

Simultaneous linear congruences

43 Find, if possible, a simultaneous solution modulo 30 to
(i) $x \equiv 1 \pmod{3}$,
$2x \equiv 4 \pmod{10}$,
(ii) $x \equiv 1 \pmod{3}$,
$2x \equiv 3 \pmod{10}$,
(iii) $x \equiv 1 \pmod{3}$,
$3x \equiv 1 \pmod{10}$,
(iv) $x \equiv 1 \pmod{3}$,
$3x \equiv 2 \pmod{10}$.

44 State conditions under which the equations

$a_1x \equiv b_1 \pmod{m_1}$,

$a_2x \equiv b_2 \pmod{m_2}$,

must have a unique solution modulo m_1m_2.

[[71]]

Lagrange's theorem for polynomials

45 Use table 3.2 to determine whether the equation $x^2 \equiv 1 \pmod{n}$ can have more than two solutions modulo n for $n = 3, 4, 5, \ldots, 16$. Why does the argument which you used in q 21 not apply to every case here?

46 Use table 3.2 to determine how many solutions there are to $x^3 \equiv 1$, modulo 5, 7, 11 and 13.
Evaluate the product $(x-1)(x-2)(x-4)$ mod 7, and the product $(x-1)(x-3)(x-9)$ mod 13.
Prove that the equation $x^2+x+1 \equiv 0$ has no solutions modulo 5 or modulo 11.

47 Use table 3.2 to determine how many solutions there are to $x^4 \equiv 1$, modulo 5, 7, 11 and 13.
Evaluate the product $(x-1)(x-2)(x-3)(x-4)$ mod 5, and the product $(x-1)(x-5)(x-8)(x-12)$ mod 13.
Factorise x^4-1, and prove that $x^2+1 \equiv 0$ has no solutions modulo 7 or modulo 11.

48 Use table 3.2 to determine how many solutions there are to $x^5 \equiv 1$, modulo 7, 11 and 13.
Evaluate the product $(x-1)(x-3)(x-4)(x-5)(x-9)$ mod 11.
Prove that $x^4+x^3+x^2+x+1 \equiv 0$ has no solutions modulo 7 or modulo 13.

Table 3.2

Powers modulo 3

0	0
1	1
2	1

Powers modulo 4

0	0	0
1	1	1
2	0	0
3	1	3

Powers modulo 5

0	0	0	0
1	1	1	1
2	4	3	1
3	4	2	1
4	1	4	1

Table 3.2 (*cont.*)

Powers modulo 6

0	0	0	0	0
1	1	1	1	1
2	4	2	4	2
3	3	3	3	3
4	4	4	4	4
5	1	5	1	5

Powers modulo 7

0	0	0	0	0	0
1	1	1	1	1	1
2	4	1	2	4	1
3	2	6	4	5	1
4	2	1	4	2	1
5	4	6	2	3	1
6	1	6	1	6	1

Powers modulo 8

0	0	0	0	0	0	0
1	1	1	1	1	1	1
2	4	0	0	0	0	0
3	1	3	1	3	1	3
4	0	0	0	0	0	0
5	1	5	1	5	1	5
6	4	0	0	0	0	0
7	1	7	1	7	1	7

Powers modulo 9

0	0	0	0	0	0	0	0
1	1	1	1	1	1	1	1
2	4	8	7	5	1	2	4
3	0	0	0	0	0	0	0
4	7	1	4	7	1	4	7
5	7	8	4	2	1	5	7
6	0	0	0	0	0	0	0
7	4	1	7	4	1	7	4
8	1	8	1	8	1	8	1

Powers modulo 10

0	0	0	0	0	0	0	0	0
1	1	1	1	1	1	1	1	1
2	4	8	6	2	4	8	6	2
3	9	7	1	3	9	7	1	3
4	6	4	6	4	6	4	6	4
5	5	5	5	5	5	5	5	5
6	6	6	6	6	6	6	6	6
7	9	3	1	7	9	3	1	7
8	4	2	6	8	4	2	6	8
9	1	9	1	9	1	9	1	9

Table 3.2 (*cont.*)

Powers modulo 11

0	0	0	0	0	0	0	0	0	0
1	1	1	1	1	1	1	1	1	1
2	4	8	5	10	9	7	3	6	1
3	9	5	4	1	3	9	5	4	1
4	5	9	3	1	4	5	9	3	1
5	3	4	9	1	5	3	4	9	1
6	3	7	9	10	5	8	4	2	1
7	5	2	3	10	4	6	9	8	1
8	9	6	4	10	3	2	5	7	1
9	4	3	5	1	9	4	3	5	1
10	1	10	1	10	1	10	1	10	1

Powers modulo 12

0	0	0	0	0	0	0	0	0	0	0
1	1	1	1	1	1	1	1	1	1	1
2	4	8	4	8	4	8	4	8	4	8
3	9	3	9	3	9	3	9	3	9	3
4	4	4	4	4	4	4	4	4	4	4
5	1	5	1	5	1	5	1	5	1	5
6	0	0	0	0	0	0	0	0	0	0
7	1	7	1	7	1	7	1	7	1	7
8	4	8	4	8	4	8	4	8	4	8
9	9	9	9	9	9	9	9	9	9	9
10	4	4	4	4	4	4	4	4	4	4
11	1	11	1	11	1	11	1	11	1	11

Powers modulo 13

0	0	0	0	0	0	0	0	0	0	0	0
1	1	1	1	1	1	1	1	1	1	1	1
2	4	8	3	6	12	11	9	5	10	7	1
3	9	1	3	9	1	3	9	1	3	9	1
4	3	12	9	10	1	4	3	12	9	10	1
5	12	8	1	5	12	8	1	5	12	8	1
6	10	8	9	2	12	7	3	5	4	11	1
7	10	5	9	11	12	6	3	8	4	2	1
8	12	5	1	8	12	5	1	8	12	5	1
9	3	1	9	3	1	9	3	1	9	3	1
10	9	12	3	4	1	10	9	12	3	4	1
11	4	5	3	7	12	2	9	8	10	6	1
12	1	12	1	12	1	12	1	12	1	12	1

Table 3.2 (*cont.*)

Powers modulo 14

0	0	0	0	0	0	0	0	0	0	0	0	0
1	1	1	1	1	1	1	1	1	1	1	1	1
2	4	8	2	4	8	2	4	8	2	4	8	2
3	9	13	11	5	1	3	9	13	11	5	1	3
4	2	8	4	2	8	4	2	8	4	2	8	4
5	11	13	9	3	1	5	11	13	9	3	1	5
6	8	6	8	6	8	6	8	6	8	6	8	6
7	7	7	7	7	7	7	7	7	7	7	7	7
8	8	8	8	8	8	8	8	8	8	8	8	8
9	11	1	9	11	1	9	11	1	9	11	1	9
10	2	6	4	12	8	10	2	6	4	12	8	10
11	9	1	11	9	1	11	9	1	11	9	1	11
12	4	6	2	10	8	12	4	6	2	10	8	12
13	1	13	1	13	1	13	1	13	1	13	1	13

Powers modulo 15

0	0	0	0	0	0	0	0	0	0	0	0	0	0
1	1	1	1	1	1	1	1	1	1	1	1	1	1
2	4	8	1	2	4	8	1	2	4	8	1	2	4
3	9	12	6	3	9	12	6	3	9	12	6	3	9
4	1	4	1	4	1	4	1	4	1	4	1	4	1
5	10	5	10	5	10	5	10	5	10	5	10	5	10
6	6	6	6	6	6	6	6	6	6	6	6	6	6
7	4	13	1	7	4	13	1	7	4	13	1	7	4
8	4	2	1	8	4	2	1	8	4	2	1	8	4
9	6	9	6	9	6	9	6	9	6	9	6	9	6
10	10	10	10	10	10	10	10	10	10	10	10	10	10
11	1	11	1	11	1	11	1	11	1	11	1	11	1
12	9	3	6	12	9	3	6	12	9	3	6	12	9
13	4	7	1	13	4	7	1	13	4	7	1	13	4
14	1	14	1	14	1	14	1	14	1	14	1	14	1

Powers modulo 16

0	0	0	0	0	0	0	0	0	0	0	0	0	0	0
1	1	1	1	1	1	1	1	1	1	1	1	1	1	1
2	4	8	0	0	0	0	0	0	0	0	0	0	0	0
3	9	11	1	3	9	11	1	3	9	11	1	3	9	11
4	0	0	0	0	0	0	0	0	0	0	0	0	0	0
5	9	13	1	5	9	13	1	5	9	13	1	5	9	13
6	4	8	0	0	0	0	0	0	0	0	0	0	0	0
7	1	7	1	7	1	7	1	7	1	7	1	7	1	7
8	0	0	0	0	0	0	0	0	0	0	0	0	0	0
9	1	9	1	9	1	9	1	9	1	9	1	9	1	9
10	4	8	0	0	0	0	0	0	0	0	0	0	0	0
11	9	3	1	11	9	3	1	11	9	3	1	11	9	3
12	0	0	0	0	0	0	0	0	0	0	0	0	0	0
13	9	5	1	13	9	5	1	13	9	5	1	13	9	5
14	4	8	0	0	0	0	0	0	0	0	0	0	0	0
15	1	15	1	15	1	15	1	15	1	15	1	15	1	15

49 Find a polynomial $g(x)$ such that $x^n - 1 = (x-1)g(x)$, where $n \geqslant 2$ is a positive integer.

50 Find a polynomial $g(x)$ such that $x^n - a^n = (x-a)g(x)$, where $n \geqslant 2$ is a positive integer.

51 If $f(x)$ is a polynomial of degree n, show that $f(x)-f(a)$ may be expressed as a product of the factor $x-a$ and a polynomial of degree $n-1$. Deduce that if $f(a)=0$, then $f(x)$ has a factor $x-a$. Prove also that if $f(x)$ is a polynomial with integral coefficients and $f(a)\equiv 0 \pmod p$, then $f(x)\equiv(x-a)g(x) \pmod p$, where $g(x)$ is a polynomial of degree $n-1$ with integral coefficients.

52 Is it possible to choose integers k, l, m, n, p, q in such a way that the polynomial

$$\begin{aligned}
&(x-1)(x-2)(x-3)(x-4)(x-5)(x-6)\\
&+k(x-1)(x-2)(x-3)(x-4)(x-5)\\
&+l(x-1)(x-2)(x-3)(x-4)\\
&+m(x-1)(x-2)(x-3)\\
&+n(x-1)(x-2)\\
&+p(x-1)\\
&+q
\end{aligned}$$

is equal to x^6-1? There is no need to actually find values. Use Fermat's theorem to prove that $x=1, 2, 3, 4, 5, 6$ all satisfy $x^6-1\equiv 0 \pmod 7$, and deduce that

$$0\equiv q\equiv p\equiv n\equiv m\equiv l\equiv k \pmod 7.$$

Deduce that $x^6-1\equiv(x-1)(x-2)(x-3)(x-4)(x-5)(x-6)$ modulo 7.

53 Can you factorise $x^{10}-1 \pmod{11}$ by working analogous to that of q 52?
Can you factorise $x^{12}-1 \pmod{13}$ by working analogous to that of q 52?
Factorise $x^{p-1}-1 \pmod p$ by working analogous to that of q 52, where p is any prime.
By looking at the number terms in this last factorisation, when p is an odd prime, give another proof of Wilson's theorem (q 25).

54 If x^3-1 has the three distinct zeros, a, b, and c modulo 19, use working like that of q 52 to prove that

$$x^3-1\equiv(x-a)(x-b)(x-c) \quad \pmod{19}.$$

Is it possible for x^3-1 to have a fourth zero, d, not congruent to a, b or c?

[[72]]

55 Prove that x^3-1 cannot have four distinct zeros modulo p, where p is a prime number.

56 If n is a positive integer, what is the maximum number of distinct zeros which x^n-1 may have modulo a prime number greater than n?

57 Use the method of the preceding questions to establish that the polynomial $a_0x^n+a_1x^{n-1}+\ldots+a_n$, with integral coefficients, $a_0 \neq 0$, cannot have more than n distinct zeros. (*Lagrange's theorem on polynomials*)

We shall use this very important theorem to establish the existence of primitive roots to a prime modulus in q 59, q 60 and q 61, and also to establish Chevalley's theorem, q 75–q 87.

Primitive roots

58 Use table 3.2 to find the orders of each of the elements in $(\mathbf{M}_p, \times)$ for $p=3, 5, 7, 11, 13$. Give a reason why the orders which occur are factors of 2, 4, 6, 10, 12 respectively. Are all the multiplicative groups $(\mathbf{M}_p, \times)$ which arise here cyclic?

59 If the group $(\mathbf{M}_p, \times)$ contains an element a of order d, what can you say about d in relation to $p-1$?
Use the element a to construct d distinct solutions to $x^d-1\equiv 0 \pmod p$. Why can there be no other solutions to this equation? Do these d solutions to $x^d-1\equiv 0 \pmod p$ form a cyclic subgroup of $(\mathbf{M}_p, \times)$? Use the fact that a cyclic group with d elements has the same structure as $(\mathbf{Z}_d, +)$ to prove that it contains $\phi(d)$ generators, and deduce that $(\mathbf{M}_p, \times)$ contains exactly $\phi(d)$ elements of order d if it contains any at all.

60 Using the symbol N_d to denote the number of elements of order d in the group $(\mathbf{M}_p, \times)$, summarise what you have learnt about the value of N_d from q 59.
If d is not a factor of $p-1$, prove that $N_d=0$.

61 If $d_1, d_2, \ldots, d_k$ are the divisors of $p-1$, including 1 and $p-1$, by counting the elements of $(\mathbf{M}_p, \times)$ by their orders, explain why

$N_{d_1}+N_{d_2}+\ldots+N_{d_k}=p-1.$

Use q 2.64 to prove that

$\phi(d_1)+\phi(d_2)+\ldots+\phi(d_k)=p-1.$

Deduce from q 60, that $N_d=\phi(d)$ for each divisor d of $p-1$, so that $N_{p-1}=\phi(p-1)$.

Since there are elements of order $p-1$ in $(\mathbf{M}_p, \times)$, it follows that $(\mathbf{M}_p, \times)$ is a cyclic group.

This is the key result on primitive roots. Questions 62–74 explore the groups $(\mathbf{M}_n, \times)$ in further detail to reveal just which of these groups are cyclic, but these results are not required in the chapters that follow. A 'primitive root' exists modulo n precisely when $(\mathbf{M}_n, \times)$ is cyclic.

62 Use table 3.2 to examine the groups $(\mathbf{M}_n, \times)$ for $n=4, 6, 8, 9, 10, 12, 14, 15, 16$. Which are cyclic and which are not? A generator of $(\mathbf{M}_n, \times)$ is called a *primitive root* modulo n.

63 Look back to q 2.47 and recall the process of deleting numbers that are not coprime to 12.

(0)	(9)	(6)	(3)
(4)	1	(10)	7
(8)	5	(2)	11

Do the numbers that remain undeleted in the column containing 1 form a reduced set of residues modulo 3?
Do the numbers that remain undeleted in the row containing 1 form a reduced set of residues modulo 4?
Are the undeleted numbers displayed as if they were part of a multiplication table modulo 12?

$1\cdot 1 \quad 1\cdot 7$

$5\cdot 1 \quad 5\cdot 7$

Is the multiplicative group $\mathbf{M}_{12}$ expressible as a direct product of the groups $\mathbf{M}_3$ and $\mathbf{M}_4$?

64 Use the table of q 2.48 to select suitable examples of the groups $(\mathbf{M}_4, \times)$ and $(\mathbf{M}_9, \times)$ which can be used to express each element of the group $\mathbf{M}_{36}$ as a product of elements of these two groups. By considering the product $19\cdot 25$, modulo 9 and modulo 4, prove, without evaluating the product, that $19\cdot 25 \equiv 7 \pmod{36}$.

65 Generalise the method of q 63 and q 64 to prove that if $\gcd(m, n)=1$, then $\mathbf{M}_{mn}=\mathbf{M}_m \times \mathbf{M}_n$.
Take particular care to prove that if $a\equiv 1 \pmod m$ and $b\equiv 1 \pmod n$, then $ab\equiv b \pmod m$ and $ab\equiv a \pmod n$.

66 Choose two suitable sets of elements from $\mathbf{M}_{63}$, one a reduced set of residues modulo 7 and one a reduced set of residues modulo 9 which would enable you to express $\mathbf{M}_{63}$ as a direct product of $\mathbf{M}_7$

and $\mathbf{M}_9$. Determine a subgroup of $\mathbf{M}_{63}$ with nine elements each of which has order 1 or 3.

67 If n is an odd number, why must $(\mathbf{M}_{2n}, \times)$ have the same structure as $(\mathbf{M}_n, \times)$?

68 The number 2 is a primitive root modulo 3, 9, 27 and 81. Use this fact to find all the primitive roots for these moduli, that is, all the generators of the groups $\mathbf{M}_3$, $\mathbf{M}_9$, $\mathbf{M}_{27}$ and $\mathbf{M}_{81}$. By using tracing paper or by making a copy of the table encircle these numbers on the array below.

1	2	3	4	5	6	7	8	9
10	11	12	13	14	15	16	17	18
19	20	21	22	23	24	25	26	27
28	29	30	31	32	33	34	35	36
37	38	39	40	41	42	43	44	45
46	47	48	49	50	51	52	53	54
55	56	57	58	59	60	61	62	63
64	65	66	67	68	69	70	71	72
73	74	75	76	77	78	79	80	81

Apparently, every primitive root modulo 9, 27, 81 is congruent to 2 modulo 3.

69 Prove that

$(3k+1)^3 \equiv 1 \pmod{9}$,

$(3k+1)^9 \equiv 1 \pmod{27}$,

$(3k+1)^{27} \equiv 1 \pmod{81}$ for all integers k,

and by induction on n deduce that $(3k+1)^{3^n} \equiv 1 \pmod{3^{n+1}}$. Deduce that no number congruent to 1 (mod 3) can be a primitive root modulo 3^n for any n.

70 If $a \in \mathbf{M}_p$, but a is not a primitive root modulo p, for an odd prime p, that is, if $a^d \equiv 1 \pmod{p}$ for $0<d<p-1$, show that

(i) $(pk+a)^d \equiv 1 \pmod{p}$,

(ii) $(pk+a)^d \equiv 1+hp \pmod{p^2}$,

(iii) $(pk+a)^{dp} \equiv 1 \pmod{p^2}$,

(iv) $(pk+a)^{dp^2} \equiv 1 \pmod{p^3}$,

where k is any integer, and h is an integer. Deduce that no number congruent to a modulo p can be a primitive root modulo p^2 or p^3.

71 When searching for primitive roots modulo 3^n among numbers of the form $3k+2$, why would either of the conditions

$(3k+2)^2 \equiv 1 \pmod{9}$ or $(3k+2)^3 \equiv 1 \pmod{9}$

make it impossible for $3k+2$ to be a primitive root modulo 9? Use Fermat's theorem to prove that $(3k+2)^3 \not\equiv 1 \pmod 9$. Deduce that the condition $(3k+2)^2 \not\equiv 1 \pmod 9$ is sufficient to ensure that $3k+2$ is a primitive root modulo 9. For which values of k is this condition satisfied?

72 If a is a primitive root modulo p, and p is an odd prime, why would either of the conditions

$$(pk+a)^{pd} \equiv 1 \pmod{p^2} \quad \text{or} \quad (pk+a)^{p-1} \equiv 1 \pmod{p^2},$$

where $0<d<p-1$, make it impossible for the number $pk+a$ to be a primitive root modulo p^2? Why, conversely, would the denial of both these conditions guarantee that $pk+a$ is a primitive root modulo p^2?
Establish that $(pk+a)^{pd} \not\equiv 1 \pmod{p^2}$.
Using the fact that

$$\begin{aligned}(a+pk)^{p-1} &= a^{p-1}+(p-1)a^{p-2}pk+p^2(\ldots)\\ &\equiv a^{p-1}-a^{p-2}pk \pmod{p^2}\\ &\equiv 1+ph-a^{p-2}pk \pmod{p^2} \text{ for some } h,\end{aligned}$$

show how to choose a value of k so that $a+pk$ is a primitive root modulo p^2.

73 Given that $(pk+a)^{p-1}=1+pu$, where u is not divisible by p, so that $pk+a$ is a primitive root modulo p^2, prove that $(pk+a)^{p(p-1)}=1+p^2v$, where v is not divisible by p, and deduce that $pk+a$ is also a primitive root modulo p^3.

74 If a is a primitive root modulo p^n, where p is an odd prime, prove that the odd number of the pair a, $a+p^n$ is a primitive root modulo $2p^n$.

In the last twelve questions of this chapter, we shall extend Lagrange's theorem on polynomials to polynomials in more than one variable and establish the existence of non-zero solutions to various polynomial equations.

Chevalley's theorem

75 Find a solution to $x^2+y^2+z^2\equiv 0$ modulo 3, modulo 5 and modulo 7, other than the solution $(x, y, z)\equiv(0, 0, 0)$.

76 If, for some prime number p, there were no solutions to $x^2+y^2+z^2\equiv 0 \pmod p$ except $(x, y, z)\equiv(0, 0, 0)$, use Fermat's theorem to prove that

$$(x^2+y^2+z^2)^{p-1}\equiv 1-(1-x^{p-1})(1-y^{p-1})(1-z^{p-1}) \pmod p$$

for all integral values of x, y and z.

[[75]]

77 The *degree* of a polynomial term $x^a y^b z^c$ is $a+b+c$. What is the highest possible degree of the terms in $(x^2+y^2+z^2)^{p-1}$? Exhibit a term of higher degree in $1-(1-x^{p-1})(1-y^{p-1})(1-z^{p-1})$.

78 Explain why any polynomial in x is equivalent modulo p to a polynomial with degree $\leqslant p-1$, in the sense that for each polynomial $f(x)$ with integral coefficients there is a polynomial $g(x)$ with integral coefficients such that $f(x) \equiv g(x) \pmod{p}$ for all integers x, and the degree of $g(x) \leqslant p-1$.

79 Prove that if two polynomials $f(x)$ and $g(x)$ with integral coefficients, both have degree $\leqslant p-1$, then $f(x) \equiv g(x) \pmod{p}$ for all integers x if and only if the coefficients of corresponding powers of x in f and g are congruent modulo p.

80 Use q 79 to prove that if $f(x, y)$ and $g(x, y)$ are polynomials with integral coefficients such that every power of x or y which appears in f or g is less than or equal to $p-1$, and $f(x, y) \equiv g(x, y) \pmod{p}$ for all integers x and y, then the coefficients of $x^i y^j$ in f and in g are congruent modulo p.

81 Explain how to extend the argument of q 80 to establish an analogous result for polynomials in three variables x, y and z.

82 By applying the result of q 81 to q 77, establish that the hypothesis of q 76 is false.

83 Establish a polynomial equivalent to $(xyz+x^2+y^2+z^2)^{p-1}$ modulo p as in q 76 on the assumption that the equation $xyz+x^2+y^2+z^2 \equiv 0 \pmod{p}$ only has the solution $(x, y, z) \equiv (0, 0, 0)$. Why is there no result analogous to q 77 in this case? (Note: Only the zero solution exists modulo 3.)

84 Modify the identity of q 76 in such a way as to construct an identity with the left hand side a polynomial of degree $2(p-1)$ and the right hand side a polynomial of degree $3(p-1)$ on the supposition that $(1, 1, 1) \pmod{p}$ is the only solution of the equation $x^2+y^2+z^2-3 \equiv 0 \pmod{p}$.

85 What conditions on the polynomial function $f(x, y, z)$ would be sufficient to apply the arguments of q 76 to show that

$$[f(x, y, z)]^{p-1} \equiv 1-(1-x^{p-1})(1-y^{p-1})(1-z^{p-1}) \pmod{p}$$

for all values of x, y and z?

⟦76⟧

86 If $[f(x, y, z)]^{p-1} \equiv 1-(1-x^{p-1})(1-y^{p-1})(1-z^{p-1}) \pmod{p}$ what constraint on the degree of the polynomial function f will lead to a contradiction of q 81?

87 Prove that every polynomial $f(x_1, x_2, \ldots, x_n)$ with integral coefficients, maximum degree $< n$, and constant term 0, has a non-trivial solution modulo p. (*Chevalley's theorem*)

Notes and answers

For concurrent reading see bibliography: Ore (1948), Shanks (1978), Davenport (1968).

1 $18-(-14)=32\equiv 0 \pmod 4$.
$b=a+4n$, $d=c+4m$,

so

$$bd=(a+4n)(c+4m)$$
$$=ac+4(am+nc+4mn)\equiv ac \pmod 4.$$

2

$\times$	-4	5	2	-1
-4	-4	-4	-4	-4
5	-4	5	2	-1
2	-4	2	-4	2
-1	-4	-1	2	5

3 $a-b=hn$, $c-d=kn$,

so

$$ac-bd=hnc+bkn=n(hc+bk)\equiv 0 \pmod n.$$

4 $a_i a_j$ belongs to a unique residue class and so is congruent to a unique a_k. If $a_Z\equiv 0$ (mod n), then $a_Z a_i\equiv 0\equiv a_Z \pmod n$. If $a_I\equiv 1 \pmod n$, then $a_I a_i\equiv a_i \pmod n$.

5 Each a_i is congruent to a unique b_i; the consequent similarity of structure is established in q 3.

6 Yes: 1.

7, 8, 9, 10 If we start with the second column, the tables give the first $n-1$ terms of the sequence.

11 $0\equiv na \pmod n$, see q 2.36.

12 4, 6, 8, 9, 10, 12, 14, 15, 16.

13 If $n=hk$, then the entry in the $(h+1)$th row and $(k+1)$th column is 0. Since each row contains n entries, a row with two zeros has an element of $\mathbf{Z}_n$ missing.

14 If p is prime, then $\gcd(a,p)=1$ for all $a\not\equiv 0 \pmod p$. From q 2.38, the sequence $a, 2a, 3a, \ldots$ does not repeat before the $(p+1)$th term. Thus the row containing a in the table has p distinct entries and so contains all of $\mathbf{Z}_p$. Thus $x\rightarrow ax$ is a permutation of $\mathbf{Z}_p$. Now $ax=ay$ in $\mathbf{Z}_p$ only if $x=y$, i.e. $x\equiv y \pmod p$. Or, $ax\equiv ay \pmod p \Rightarrow a(x-y)\equiv 0 \pmod p \Rightarrow p|a$ or $p|x-y$. Also if $x\rightarrow ax$ is a permutation of $\mathbf{Z}_p$, then for any $b\in\mathbf{Z}_p$, there is a unique x such that $ax=b$.

15 Under $x \to ax$, $0 \to 0$, so $x \to ax$ is a permutation of the non-zero elements of $\mathbf{Z}_p$. Thus $\{1, 2, \ldots, p-1\} \equiv \{a, 2a, \ldots, (p-1)a\}$ (mod p), and
$$1 \cdot 2 \ldots (p-1) \equiv a \cdot 2a \ldots (p-1)a \pmod{p}.$$
Cancelling as in q 14, we have $1 \equiv a^{p-1}$ (mod p).

16 Put $b = 1$, or $a^{p-2}a \equiv 1$ (mod p).

17 If neither a nor b has a factor p, nor does ab, so the set is closed; identity 1; inverses established in q 16. The group of non-zero residues modulo p under multiplication may be expressed either by multiplication modulo p on a complete set of residues with the residue congruent to zero deleted, or by taking the non-zero residue classes as elements and using the multiplication induced by the ordinary multiplication on the integers.

The existence of this group is fundamental to the rest of this chapter. We shall eventually establish, in q 61, that this group must be cyclic.

18 Non-zero elements of $\mathbf{Z}_p$ form a group of order $p-1$. By Lagrange's theorem, the order of each element divides the order of the group, so $x^d = 1$ in the group for some $d|p-1$. Now $x^{p-1} = x^{d(p-1)/d} = 1^{(p-1)/d} = 1$.

19 If $x \not\equiv 0$ (mod p), then $x^{p-1} \equiv 1 \Rightarrow x^p \equiv x$.
If $x \equiv 0$ (mod p), then $x^p \equiv 0 \equiv x$.

The results of q 15, q 18 and q 19 are worth remembering. They will be needed from time to time in what follows.

20 $2222 \equiv 3$ (mod 7), $5555 = 4$ (mod 7), $5555 \equiv 5$ (mod 6), $2222 \equiv 2$ (mod 6).
$2222^{5555} + 5555^{2222} \equiv 3^5 + 4^2 \equiv 12 + 2 \equiv 0$ (mod 7).

21 ± 1. If $x^2 - 1 \equiv 0$ (mod p), then $p|x-1$ or $p|x+1$.

22 $1 \cdot 2 \cdot 3 \cdot 4 \cdot 5 \cdot 6 = (2 \cdot 4)(3 \cdot 5)6 \equiv 6$ (mod 7).

23 $1 \cdot 2 \cdot 3 \cdot 4 \cdot 5 \cdot 6 \cdot 7 \cdot 8 \cdot 9 \cdot 10 = (2 \cdot 6)(3 \cdot 4)(5 \cdot 9)(7 \cdot 8)10 \equiv 10 \equiv -1$ (mod 11).

24 All except for those satisfying $x^2 = 1$.

25 $(p-1)! \equiv -1$ (mod p).

Wilson's theorem is derived here more for its intrinsic interest than for its applications.

26 6, 13; none; 8 only.

27 (i) 0, 2, 4, 6, 8, 10, 12,
(ii) 1, 3, 5, 7, 9, 11, 13.

28 $4x \equiv b \pmod{14} \Rightarrow b = 4x + 14k$, so b is even.

29 $2c = 4x + 14k \Leftrightarrow c = 2x + 7k \Leftrightarrow 2x \equiv c \pmod 7$. $2 \cdot 4 = 1 \pmod 7$ so $2x \equiv c \pmod 7 \Leftrightarrow x \equiv 4c \pmod 7$. Of the fourteen numbers $4c, 4c+1, \ldots, 4c+13$, just $4c$ and $4c+7$ are congruent to $4c \pmod 7$.

30 $ax \equiv b \pmod n \Rightarrow b = ax + kn \Rightarrow \gcd(a, n) | b$.

31 $ar + ns = 1 \Rightarrow arb + nsb = b \Rightarrow arb \equiv b \pmod n$, so $ax \equiv b \pmod n$ has a solution.

32 $ax \equiv b \pmod n$ and $ay \equiv b \pmod n \Rightarrow ax \equiv ay \pmod n$, so $n | a(x-y)$. But $\gcd(a, n) = 1$, so $n | x - y$ and $x \equiv y \pmod n$.

33 If $\gcd(a, n) = d$, then $\gcd(a/d, n/d) = 1$. Now $ax \equiv b \pmod n \Rightarrow b = ax + kn$ and $d | b \Rightarrow b/d = (a/d)x + k(n/d)$ is an equation in integers, so $(a/d)x \equiv b/d \pmod{n/d}$ and this has a solution by q 31, which is unique for $0 \leq x < n/d$ by q 32. If r is this solution, then of the numbers $r, r+1, \ldots, r+(n-1)$, only those of the form $r + k(n/d)$ are congruent to $r \pmod{n/d}$, so we obtain distinct solutions of $ax \equiv b \pmod n$ for $k = 0, 1, \ldots, d-1$.
In fact, every integer x satisfying $ax \equiv b \pmod n$ satisfies $(a/d)x \equiv b/d \pmod{n/d}$ and conversely.

34 $a = 1, 3, 5, 9, 11, 13$.
If $x \to ax$ is a permutation, then $ax \equiv 1$ for some x.
If $ax \equiv 1$, then $a(xb) \equiv b$, so $x \to ax$ is an onto map.

35 $a = 1, 5, 7, 11$.

36 The equivalence of (ii) and (iii) is proved in n 34.
We establish equivalence of (i) and (ii) by

$$\gcd(a, n) = 1 \Leftrightarrow \text{there exist integers } x, y \text{ with } ax + ny = 1$$
$$\Leftrightarrow ax \equiv 1 \pmod n.$$

From (i), there are $\phi(n)$ elements in a reduced set of residues.
$\mathbf{M}_n$ can also be defined as the set of generators of $(\mathbf{Z}_n, +)$.

37 $\gcd(a, p) = 1 \Leftrightarrow a \not\equiv 0 \pmod p$.

It is important to retain the freedom to think of $\mathbf{M}_n$ as consisting either of a reduced set of residues with multiplication modulo n, or of the corresponding residue classes with the naturally induced multiplication as in q 3.

38

1	3
3	1

1	5
5	1

1	3	5	7
3	1	7	5
5	7	1	3
7	5	3	1

1	2	4	5	7	8
2	4	8	1	5	7
4	8	7	2	1	5
5	1	2	7	8	4
7	5	1	8	4	2
8	7	5	4	2	1

1	3	7	9
3	9	1	7
7	1	9	3
9	7	3	1

39 Yes.

40 Plainly $1 \in \mathbf{M}_n$. Since $ab \equiv 1 \Rightarrow ba \equiv 1$, each element in $\mathbf{M}_n$ has an inverse. If $ab \equiv 1$ and $cd \equiv 1$, then $(ac)(bd) \equiv 1$, so $\mathbf{M}_n$ is closed.

41 Because $\mathbf{M}_n$ is a group, $x \to a_i x$ is a permutation of $\mathbf{M}_n$, so $\{a_1, \ldots, a_{\phi(n)}\} \equiv \{a_i a_1, \ldots, a_i a_{\phi(n)}\} \pmod{n}$. Now $a_1 \ldots a_{\phi(n)} \equiv a_i a_1 \ldots a_i a_{\phi(n)} \pmod{n}$. By cancellation $1 \equiv (a_i)^{\phi(n)} \pmod{n}$.

42 Since $\mathbf{M}_n$ is a group of order $\phi(n)$, the order of each element divides $\phi(n)$, so that for $a \in \mathbf{M}_n$, $a^{\phi(n)} = 1$.

43 (i) 7, 22, (ii) none, (iii) 7, (iv) 4.

44 From q 31 and q 32, $\gcd(a_1, m_1) = \gcd(a_2, m_2) = 1$ gives a unique solution to each equation, and then $\gcd(m_1, m_2) = 1$ gives a unique solution to the pair, by the Chinese remainder theorem.

45 When $n = 8$, 12, 15 or 16, there are four solutions to $x^2 = 1$. Although $x^2 \equiv 1 \pmod{n} \Rightarrow n | (x+1)(x-1)$, when n is composite we need not have either $n | x+1$ or $n | x-1$.

46 Mod 5; 1. Mod 7; 1, 2, 4. Mod 11; 1. Mod 13; 1, 3, 9.

$(x-1)(x-2)(x-4) \equiv x^3 - 1 \pmod{7}$,

$(x-1)(x-3)(x-9) \equiv x^3 - 1 \pmod{13}$.

$x^3 - 1 = (x-1)(x^2+x+1)$, so a solution of $x^2+x+1 \equiv 0$ must be a solution of $x^3 - 1 \equiv 0$.

47 Mod 5; 1, 2, 3, 4, by Fermat's theorem.
Mod 7; 1, 6. Mod 11; 1, 10. Mod 13; 1, 5, 8, 12.

$(x-1)(x-2)(x-3)(x-4) \equiv x^4 - 1 \pmod{5}$,

$(x-1)(x-5)(x-8)(x-12) \equiv x^4 - 1 \pmod{13}$.

$x^4 - 1 = (x-1)(x+1)(x^2+1)$, so a solution of $x^2+1 \equiv 0$ must be a solution of $x^4 - 1 \equiv 0$.

48 Mod 7; 1. Mod 11; 1, 3, 4, 5, 9. Mod 13; 1.

$(x-1)(x-3)(x-4)(x-5)(x-9) \equiv x^5 - 1 \pmod{11}$.

$x^5 - 1 = (x-1)(x^4+x^3+x^2+x+1)$, so a solution of $x^4+x^3+x^2+x+1 \equiv 0$ must be a solution of $x^5 - 1 \equiv 0$.

49 $g(x) = x^{n-1} + x^{n-2} + \ldots + x + 1$.

50 $g(x) = x^{n-1} + ax^{n-2} + \ldots + a^i x^{n-i-1} + \ldots + a^{n-1}$.

51 If $f(x) = b_n x^n + b_{n-1} x^{n-1} + \ldots + b_0$, then

$f(x) - f(a) = b_n(x^n - a^n) + b_{n-1}(x^{n-1} - a^{n-1}) + \ldots + b_1(x-a)$

and from q 50 each term has a factor $x - a$.
So $f(x) - f(a) = (x-a)g(x)$, and if $f(a) = 0$, $f(x) = (x-a)g(x)$.

Also $f(x) = (x-a)g(x) + f(a)$, so if $f(a) \equiv 0 \pmod p$,

$f(x) \equiv (x-a)g(x) \pmod p$.

52 Choose k to make the coefficient of $x^5 = 0$.
Choose l to make the coefficient of $x^4 = 0$.
Choose m to make the coefficient of $x^3 = 0$.
Choose n to make the coefficient of $x^2 = 0$.
Choose p to make the coefficient of $x = 0$.
Choose q to make the number term -1.

53 $x^{p-1} - 1 \equiv (x-1)(x-2)\ldots(x-(p-1)) \pmod p$. $p-1$ is an even number if p is odd and the constant term on the right is $(p-1)!$. If $p = 2$, Wilson's theorem is trivial.

54 Let $x^3 - 1 = (x-a)(x-b)(x-c) + k(x-a)(x-b) + l(x-a) + m$.
Then since $a^3 - 1 \equiv 0 \pmod{19}$, $m \equiv 0 \pmod{19}$.
Since $b^3 - 1 \equiv 0 \pmod{19}$, $b \not\equiv a \pmod{19}$, $l \equiv 0 \pmod{19}$.
Since $c^3 - 1 \equiv 0 \pmod{19}$, $c \not\equiv a, b \pmod{19}$, $k \equiv 0 \pmod{19}$.
So $x^3 - 1 \equiv (x-a)(x-b)(x-c) \pmod{19}$.
Now $0 \equiv d^3 - 1 \equiv (d-a)(d-b)(d-c) \pmod{19}$.
But this implies $d \equiv a$, b, or $c \pmod{19}$.

55 As n 54 with p for 19.

56 Just n; argument as n 54.

57 If $b_1, \ldots, b_n$ are distinct zeros, we can use the method of q 52 to show that

$a_0x^n + \ldots + a_n = a_0(x-b_1)\ldots(x-b_n)$.

Now a further zero of the polynomial on the left must give at least one zero among the factors on the right. The result holds whether we are working over the integers or modulo a prime number; in this latter case we must have $a_0 \not\equiv 0 \pmod p$.

58

Mod 3	element	1	2									
	order	1	2									
Mod 5	element	1	2	3	4							
	order	1	4	4	2							
Mod 7	element	1	2	3	4	5	6					
	order	1	3	6	3	6	2					
Mod 11	element	1	2	3	4	5	6	7	8	9	10	
	order	1	10	5	5	5	10	10	10	5	2	
Mod 13	element	1	2	3	4	5	6	7	8	9	10	11	12
	order	1	12	3	6	4	12	12	4	3	6	12	2

The groups here have orders 2, 4, 6, 10, 12 respectively. These groups are cyclic as we can find a generator in each case.

59 By Lagrange's theorem $d \mid p-1$.
$1, a, a^2, \ldots, a^{d-1}$ are distinct solutions. By Lagrange's theorem, q 57, these are all the solutions. So every element of order d is in this set. The

element of order d are just the generators of this group, so there are $\phi(d)$ of them.

60 Either $N_d = 0$, or $N_d = \phi(d)$.
If d is not a factor of $p-1$, $\mathbf{M}_p$ contains no elements of order d by Lagrange's theorem on subgroups.

61 Every element has an order and so must be counted once. The orders are all divisors of $p-1$. Since $N_{d_i} \leq \phi(d_i)$ and both the sums are the same, $N_{d_i} = \phi(d_i)$ for each divisor.

62 Cyclic for $n = 4, 6, 9, 10, 14$. We have proved so far that there exists a primitive root when n is prime, and when $n = 4, 6, 9, 10, 14$.

63 Each column is a complete set of residues modulo 3 and each row is a complete set of residues modulo 4. The deletions effect the reduction. Working modulo 12, $\{1, 5\} \cdot \{1, 7\} = \{1, 5, 7, 11\}$, or $\mathbf{M}_3 \cdot \mathbf{M}_4 = \mathbf{M}_{12}$.

64 Mod 36

	1	29	13	5	25	17
	19	11	31	23	7	35

Mod 9 (across); Mod 4 (down)

$19 \cdot 25 \equiv 1 \cdot 7 \equiv 7 \pmod 9$; $19 \cdot 25 \equiv 3 \cdot 1 \equiv 3 \pmod 4$, and by the Chinese remainder theorem, there is just one number modulo 36 which is congruent to 7 (mod 9) and 3 (mod 4). 7 is such a number.
$\{1, 29, 13, 5, 25, 17\}$ is a reduced set of residues modulo 9, and each element $\equiv 1 \pmod 4$, thus it forms a subgroup like $\mathbf{M}_9$ in $\mathbf{M}_{36}$. Likewise, $\{1, 19\}$ is a reduced set of residues modulo 4, and each element $\equiv 1 \pmod 9$ so it forms a subgroup like $\mathbf{M}_4$ in $\mathbf{M}_{36}$.

65 From q 2.50, the array which remains when the numbers not coprime to mn are deleted has $\phi(m)$ rows and $\phi(n)$ columns. The row containing 1 is a reduced set of residues modulo n, each element in it is congruent to $1 \pmod m$, so the row is a subgroup like $\mathbf{M}_n$ in $\mathbf{M}_{mn}$. Similarly the column containing 1 is a subgroup like $\mathbf{M}_m$ in $\mathbf{M}_{mn}$. Now if $b \equiv 1 \pmod n$, then $ab \equiv a \pmod n$, so a and ab lie in the same column. If $a \equiv 1 \pmod m$, then $ab \equiv b \pmod m$, so b and ab lie in the same row. Thus the array displays the direct product $\mathbf{M}_{mn} = \mathbf{M}_m \cdot \mathbf{M}_n$.

66 In $\mathbf{M}_{63}$, $\{1, 10, 19, 37, 46, 55\} = \mathbf{M}_7$, and $\{1, 8, 22, 29, 43, 50\} = \mathbf{M}_9$. Every element in the square array, other than 1, has order 3 in $\mathbf{M}_{63}$.

1	37	46
22	58	4
43	16	25

This is the direct product of two groups of order 3.

67 If n is odd, then $\gcd(2, n) = 1$, so $\mathbf{M}_{2n} = \mathbf{M}_2 \cdot \mathbf{M}_n$, but $\mathbf{M}_2 = \{1\}$.

68 Mod 9; 2, $2^5=5$.
Mod 27; 2, $2^5=5$, $2^7=20$, $2^{11}=23$, $2^{13}=11$, $2^{17}=14$.
Mod 81; 2, $2^5=32$, $2^7=47$, $2^{11}=23$, $2^{13}=11$, $2^{17}=14$, $2^{19}=56$, $2^{23}=5$, $2^{25}=20$, $2^{29}=77$, $2^{31}=65$, $2^{35}=68$, $2^{37}=29$, $2^{41}=59$, $2^{43}=74$, $2^{47}=50$, $2^{49}=38$, $2^{53}=41$.

69 $(3k+1)^3=27k^3+27k^2+9k+1\equiv 1 \pmod 9$,
$(3k+1)^9=(9h+1)^3=9^3h^3+3\cdot 9^2h^2+27h+1\equiv 1 \pmod{27}$,
$(3k+1)^{27}=(27g+1)^3=27^3g^3+3\cdot 27g^2+81g+1\equiv 1 \pmod{81}$.
If $(3k+1)^{3^{n-1}}\equiv 1 \pmod{3^n}$, then

$$(3k+1)^{3^n}=(3^nh+1)^3\equiv 1 \pmod{3^{n+1}}.$$

70 (i) by q 3,
(ii) from the meaning of (i),
(iii) $(pk+a)^{dp}=(1+hp)^p\equiv 1 \pmod{p^2}$.
(iv) $(pk+a)^{dp^2}=(1+gp^2)^p\equiv 1 \pmod{p^3}$.
Thus $pk+a$ does not have order $p(p-1)$ modulo p^2 or order $p^2(p-1)$ modulo p^3.

71 The order of $3k+2$ would then be 1, 2 or 3, not 6. $(3k+2)^3\equiv 2^3 \pmod 3$. $2^3\not\equiv 1 \pmod 3$. Thus $3k+2$ is not of order 3 (mod 9). If $(3k+2)^2\not\equiv 1 \pmod 9$, then $3k+2$ is not of order 1 or 2, so this condition implies that $3k+2$ has order 6.

$$\begin{aligned} 9k^2+12k+4\not\equiv 1 \pmod 9 &\Leftrightarrow 3k+3\not\equiv 0 \pmod 9\\ &\Leftrightarrow k+1\not\equiv 0 \pmod 3\\ &\Leftrightarrow k\equiv 0, 1 \pmod 3. \end{aligned}$$

72 A primitive root modulo p^2 has order $p(p-1)$. Either of the given conditions contradicts this. If both are false then no proper factor of $p(p-1)$ can be the order of $pk+a$.

$$\begin{aligned} (pk+a)^{pd} &\equiv (pk+a)^d \pmod p \quad \text{by Fermat's theorem}\\ &\equiv a^d \pmod p\\ &\not\equiv 1 \pmod p \text{ since } a \text{ is a primitive root and } d<p-1. \end{aligned}$$

Choose k so that $h-a^{p-2}k\not\equiv 0 \pmod p$.

73 $(pk+a)^{p-1}=1+pu\Rightarrow (pk+a)^{p(p-1)}=(1+pu)^p=1+p^2u+p^3(\ldots)$.
Also $(pk+a)^{dp^2}\equiv a^d\not\equiv 1 \pmod p$ if $d<p-1$.

74 The order of a primitive root modulo p^n is the same as the order of a primitive root modulo $2p^n$, namely $p^{n-1}(p-1)$. Thus $(a+kp^n)$ belongs to $\mathbf{M}_{2p^n}$ if it is odd and necessarily has the right order for a primitive root.

75 (1, 1, 1) (mod 3), (0, 1, 2) (mod 5), (1, 2, 3,) (mod 7).

76 By the hypothesis and Fermat's theorem, the left hand side $\equiv 1$ unless $x\equiv y\equiv z\equiv 0$. If one of $x, y, z\not\equiv 0$, then $(1-x^{p-1})(1-y^{p-1})(1-z^{p-1})\equiv 0$ by Fermat's theorem.

77 $2(p-1)$; $x^{p-1}y^{p-1}z^{p-1}$ has degree $3(p-1)$.

78 $x^p \equiv x \pmod p$ so $x^{p+n} \equiv x^{1+n} \pmod p$.

79 $f(x)-g(x) \equiv 0 \pmod p$ for distinct non-congruent integers, but the degree of $f(x)-g(x) \leqslant p-1$, so this contradicts Lagrange's theorem, q 57. Thus $f(x)-g(x)$ must be the zero polynomial modulo p, and coefficients of the same powers of x in f and g are congruent modulo p.

80 Regard $f(x, y)$ and $g(x, y)$ as polynomials in x, and let the coefficients of x^i in f and g be $a_f(y)$ and $a_g(y)$ respectively, then by q 79, $a_f(y) \equiv a_g(y)$ for all integers y, and applying q 79 again, the coefficients of the same powers of y in a_f and a_g are congruent modulo p.

81 Regard $f(x, y, z)$ and $g(x, y, z)$ as polynomials in z. Then use q 79 and q 80.

82 Coefficient of $x^{p-1}y^{p-1}z^{p-1}$ on the right is 1, and on the left is 0. But these must be congruent modulo p by q 81.

83 $(xyz+x^2+y^2+z^2)^{p-1} \equiv 1-(1-x^{p-1})(1-y^{p-1})(1-z^{p-1})$. Both left and right have coefficient 1 for $x^{p-1}y^{p-1}z^{p-1}$.

84 $(x^2+y^2+z^2-3)^{p-1}$

$$\equiv 1-[1-(x-1)^{p-1}][1-(y-1)^{p-1}][1-(z-1)^{p-1}]$$

85 $f(x, y, z) \equiv 0$ only when $x \equiv y \equiv z \equiv 0 \pmod p$.

86 Maximum degree of $f(x, y, z) < 3$.

87 If $f(x_1, \ldots, x_n) \equiv 0$ only when $x_1 \equiv \ldots \equiv x_n \equiv 0 \pmod p$, then $[f(x_1, \ldots, x_n)]^{p-1} \equiv 1-(1-x_1^{p-1}) \ldots (1-x_n^{p-1})$.
If the maximum degree of f is less than n, this contradicts the generalisation of q 81 to n variables.

Historical note

Fermat's theorem is claimed in a letter which he wrote in 1640. In 1773, J. L. Lagrange was the first to publish a proof of the theorem which had been ascribed to Wilson some three years earlier. Euler's generalisation of Fermat's theorem was published in 1760. Using Lagrange's work on polynomials, A. M. Legendre gave a theoretical construction for a primitive root modulo p in 1785. C. F. Gauss (1801) showed that primitive roots existed just for the moduli 2, 4, p^n and $2p^n$. The discussion of such properties in terms of the group of units of a quotient ring belongs to the twentieth century. What we have called Chevalley's theorem is a simple version of some work published by C. Chevalley in 1936.

A careful survey of most of the results in the first three chapters of this book is given in Ore (1948).

4

Quadratic residues

Quadratic residues and the Legendre symbol

1 Working in $\mathbf{Z}_7$, find all the solutions to $x^2 = 0, 1, 2, 3, 4, 5, 6$ respectively.

2 In $\mathbf{Z}_7$, which of the equations

(i) $(x+2)^2 = 0$,	(v) $x^2 + 2x = 3$,	(ix) $x^2 + 3x = 3$,
(ii) $(x+2)^2 = 1$,	(vi) $x^2 + 2x = 4$,	(x) $x^2 + 3x = 4$,
(iii) $(x+2)^2 = 2$,	(vii) $x^2 + 2x = 5$,	(xi) $x^2 + 3x = 5$,
(iv) $(x+2)^2 = 3$,	(viii) $x^2 + 2x = 6$,	(xii) $x^2 + 3x = 6$,

have no solution, have one solution, have two solutions?

3 The perfect squares in $\mathbf{M}_7$ are called the *quadratic residues* modulo 7. Exhibit the mapping of $\mathbf{M}_7$ given by $x \to x^2$. Does the set of images under this mapping form a subgroup of $\mathbf{M}_7$?

4 Use table 3.2 to find the quadratic residues modulo 5, 11, 12 and 13. For which of these moduli do the quadratic residues form a subgroup of $\mathbf{M}_n$? For which of these values of n is the mapping of $\mathbf{M}_n$ given by $x \to x^2$, two to one?

5 State the number of quadratic residues modulo 3, 5, 7, 11, 13 and 17 respectively. Predict the number of quadratic residues modulo p (an odd prime).

6 If $x^2 \equiv y^2 \pmod p$, does it follow that either $x \equiv y \pmod p$ or $x \equiv -y \pmod p$, when p is a prime number?

7 For any prime p, determine the elements of $\mathbf{M}_p$ which are mapped to 1 under the mapping $x \to x^2$. Can you say how many elements of $\mathbf{M}_p$ are mapped to any other quadratic residue under this mapping?

8 A number in $\mathbf{M}_n$ which is not a square is called a *quadratic non-residue* modulo n. Find three quadratic non-residues modulo 12.

9 Find the cube of all the quadratic residues modulo 7, and also the cube of all the quadratic non-residues.

10 What does Fermat's theorem imply about the number of roots of $x^6 \equiv 1 \pmod 7$?
What can you deduce about the cube of a quadratic residue modulo 7?
By factorising $x^6 - 1$ and using Lagrange's theorem, q 3.57, deduce that at most three cubes are congruent to 1 (mod 7) and at most three cubes are congruent to -1 (mod 7).

11 Using the method of q 10, determine how many fifth powers are congruent to 1 (mod 11) and how many fifth powers are congruent to -1 (mod 11). Does the value of the fifth power determine whether the number is a quadratic residue or not?

12 Generalise the method of q 10 and q 11 to give a rule for determining which are the quadratic residues and which the quadratic non-residues modulo p (an odd prime).

13 Use the fact that $(xy)^k = x^k y^k$ to show that the set of kth powers in $\mathbf{M}_n$ form a subgroup.

14 If p is an odd prime, what is the image of $\mathbf{M}_p$ under the mapping $x \to x^{\frac{1}{2}(p-1)}$?

15 The *Legendre symbol* $\left(\frac{a}{p}\right)$, or sometimes $(a|p)$, is conventionally used to denote +1 when a is a quadratic residue modulo p, -1 when a is a quadratic non-residue modulo p, and 0 otherwise. Explain why

$$\left(\frac{ab}{p}\right) = \left(\frac{a}{p}\right)\left(\frac{b}{p}\right).$$

16 List those prime numbers less than 20 for which
(i) -1 is a quadratic residue,
(ii) -1 is a quadratic non-residue.
Is there a pattern which you can see?

17 What is the order of -1 as an element of the group $\mathbf{M}_p$? Can any other element of $\mathbf{M}_p$ have the same order as -1?

18 What can you say about the odd prime number p
(i) if $(-1)^{\frac{1}{2}(p-1)} \equiv 1 \pmod p$,
(ii) if $(-1)^{\frac{1}{2}(p-1)} \equiv -1 \pmod p$?

⟦86⟧

19 What is the order of the subgroup of $\mathbf{M}_p$ consisting of quadratic residues? For what values of p is this order even? For what values of p is this order odd? Suppose here that p is an odd prime number.

20 Why is it impossible for a group of odd order to contain an element of order 2?

21 By pairing elements with their inverses explain why a group of even order must contain an element of order 2.

22 If p is a prime number of the form $4k+1$, must $\mathbf{M}_p$ contain an element of order 4?

23 Find the odd prime factors of the numbers 2^2+1, 3^2+1, 4^2+1, 5^2+1, 6^2+1, 7^2+1, 8^2+1, 9^2+1.
How many of the prime factors have the form $4k+1$, and how many have the form $4k+3$?

24 If p is an odd prime which divides a number of the form n^2+1, can you prove that p has the form $4k+1$?

25 Modify the approach of q 1.65 to prove that there is an infinity of primes of the form $4k+1$.

26 Find the prime factors of 2^2+2+1, 3^2+3+1, 4^2+4+1, 5^2+5+1, 6^2+6+1, 7^2+7+1, 8^2+8+1, 9^2+9+1.
How many of the prime factors have the form $3k+1$?

27 If p is an odd prime which divides a number of the form n^2+n+1, prove that $p=3$ or p has the form $3k+1$, since $\mathbf{M}_p$ contains an element of order 3.

28 Modify the approach of q 25 to prove that there is an infinity of primes of the form $3k+1$.

29 If p is a prime number of the form $4k+1$, explain how Wilson's theorem, q 3.25, implies that

$$(1\cdot 2\cdot 3\ldots 2k)^2 \equiv -1 \pmod{p}.$$

Gauss' lemma

The next thirteen questions gradually develop a tool which will help to determine whether a number is a quadratic residue modulo p or not.

30 In table 4.1, multiplication tables for $\mathbf{M}_3$, $\mathbf{M}_5$, $\mathbf{M}_7$, $\mathbf{M}_{11}$ and $\mathbf{M}_{13}$ are given using the smallest absolute values of the residues. Explain why the first half of the first row of each table is repeated

Table 4.1

Multiplication modulo 3 numerically least residues

1	−1
−1	1

Multiplication modulo 5 numerically least residues

1	2	−2	−1
2	−1	1	−2
−2	1	−1	2
−1	−2	2	1

Multiplication modulo 7 numerically least residues

1	2	3	−3	−2	−1
2	−3	−1	1	3	−2
3	−1	2	−2	1	−3
−3	1	−2	2	−1	3
−2	3	1	−1	−3	2
−1	−2	−3	3	2	1

Multiplication modulo 11 numerically least residues

1	2	3	4	5	−5	−4	−3	−2	−1
2	4	−5	−3	−1	1	3	5	−4	−2
3	−5	−2	1	4	−4	−1	2	5	−3
4	−3	1	5	−2	2	−5	−1	3	−4
5	−1	4	−2	3	−3	2	−4	1	−5
−5	1	−4	2	−3	3	−2	4	−1	5
−4	3	−1	−5	2	−2	5	1	−3	4
−3	5	2	−1	−4	4	1	−2	−5	3
−2	−4	5	3	1	−1	−3	−5	4	2
−1	−2	−3	−4	−5	5	4	3	2	1

Multiplication modulo 13 numerically least residues

1	2	3	4	5	6	−6	−5	−4	−3	−2	−1
2	4	6	−5	−3	−1	1	3	5	−6	−4	−2
3	6	−4	−1	2	5	−5	−2	1	4	−6	−3
4	−5	−1	3	−6	−2	2	6	−3	1	5	−4
5	−3	2	−6	−1	4	−4	1	6	−2	3	−5
6	−1	5	−2	4	−3	3	−4	2	−5	1	−6
−6	1	−5	2	−4	3	−3	4	−2	5	−1	6
−5	3	−2	6	1	−4	4	−1	−6	2	−3	5
−4	5	1	−3	6	2	−2	−6	3	−1	−5	4
−3	−6	4	1	−2	−5	5	2	−1	−4	6	3
−2	−4	−6	5	3	1	−1	−3	−5	6	4	2
−1	−2	−3	−4	−5	−6	6	5	4	3	2	1

in reverse order in the second half of the first row, if only absolute values are considered. Infer why the same is true of each column of these tables. Deduce that all possible absolute values appear in the first half of each row.

31 In table 4.1, count the minus signs in the first half of each row. Make a list of the elements of $\mathbf{M}_p$ for $p=3, 5, 7, 11, 13$ and record whether the corresponding number of minus signs which you have counted is odd or even. Compare the list you have made with a list of quadratic residues and non-residues.

32 Express each of the numbers

$2\cdot 1, 2\cdot 2, 2\cdot 3, 2\cdot 4, 2\cdot 5, 2\cdot 6, 2\cdot 7, 2\cdot 8$

as a numerically least residue modulo 17, that is a number between +8 and −8. Deduce that $2^8\equiv(-1)^4\equiv 1 \pmod{17}$. Use q 12 to determine whether 2 is a quadratic residue modulo 17 or not.

33 Modulo 17, we have

$3\cdot 1\equiv 3,\ 3\cdot 2\equiv 6,\ 3\cdot 3\equiv -8,\ 3\cdot 4\equiv -5,$

$3\cdot 5\equiv -2,\ 3\cdot 6\equiv 1,$

$3\cdot 7\equiv 4,\ 3\cdot 8\equiv 7.$

Deduce that $3^8\equiv -1 \pmod{17}$, and determine whether 3 is a quadratic residue modulo 17 or not.

34 Use the method of q 32 to determine whether 2 is a quadratic residue modulo 19 or not.

35 If a is an element of $\mathbf{M}_{17}$, explain why the numerically least residues congruent to

$a\cdot 1, a\cdot 2, a\cdot 3, a\cdot 4, a\cdot 5, a\cdot 6, a\cdot 7, a\cdot 8$

have distinct absolute values. Show how to determine $a^8 \pmod{17}$, and to decide whether a is a quadratic residue.

36 If p is an odd prime number, explain why the numerically least residues congruent to

$2\cdot 1, 2\cdot 2, 2\cdot 3, \ldots, 2\cdot \frac{1}{2}(p-1)$ modulo p

have distinct absolute values. How would you determine the value of $2^{\frac{1}{2}(p-1)} \pmod{p}$?

37 Use q 36 to prove that if $p=8k+1$, then 2 is a quadratic residue modulo p.

⟦89⟧

38 Use q 36 to prove that if $p=8k+3$, then 2 is a quadratic non-residue modulo p.

39 Use q 36 to find out whether 2 is a quadratic residue modulo $p=8k+5$.

40 Use q 36 to find out whether 2 is a quadratic residue modulo $p=8k+7$.

41 Determine whether $\frac{1}{8}(p^2-1)$ is odd or even, when $p=8k+1$, $8k+3$, $8k+5$, $8k+7$, and deduce that $\left(\frac{2}{p}\right)=(-1)^{(p^2-1)/8}$.

Although this result takes a rather odd form it is one of the basic results needed for the general determination of quadratic residues. Because it is not easy to remember, you should remember where to look it up.

42 Determine the values of $\left(\frac{-2}{p}\right)$ when $p=8k+1$, $8k+3$, $8k+5$ and $8k+7$.

43 If p is an odd prime, and a is an element of $\mathbf{M}_p$, explain why the numerically least residues congruent to

$a\cdot 1, a\cdot 2, a\cdot 3, \ldots, a\cdot \frac{1}{2}(p-1)$

have distinct absolute values. Prove that if there are l minus signs in this list then $a^{\frac{1}{2}(p-1)}=(-1)^l$. (*Gauss' lemma*)

Law of quadratic reciprocity

The next question introduces one new piece of notation.

44 We let $[\frac{7}{2}]=3$, $[3]=3$, $[\sqrt{2}]=1$, $[0.9]=0$, $[-\frac{1}{2}]=-1$, so that when x is a real number,

$[x]$ is an integer,

$[x]\leqslant x<[x]+1$,

and $[x]$ is called the integral part of x.
Find $\frac{11}{13}-[\frac{11}{13}]$, $\frac{22}{13}-[\frac{22}{13}]$, $\frac{33}{13}-[\frac{33}{13}]$, $\frac{44}{13}-[\frac{44}{13}]$, $\frac{55}{13}-[\frac{55}{13}]$, $\frac{66}{13}-[\frac{66}{13}]$. How many of these numbers exceed $\frac{1}{2}$?

45 If the numerically least residue congruent to x (mod 13) is negative, what can be said about the value of $x/13-[x/13]$?
If the numerically least residue congruent to x (mod 13) is positive, what can be said about the value of $x/13-[x/13]$?

46 From q 43, q 44 and q 45, determine whether 11 is a quadratic residue modulo 13, and check your answer with table 3.2.

⟦89⟧

47 Using squared paper with grid lines at least a centimetre apart, or similar lattice point paper, mark the points

$$\left(1, \frac{11}{13}\right), \left(2, \frac{22}{13}\right), \left(3, \frac{33}{13}\right), \left(4, \frac{44}{13}\right), \left(5, \frac{55}{13}\right), \left(6, \frac{66}{13}\right)$$

and the points

$$\left(1, \left[\frac{11}{13}\right]\right), \left(2, \left[\frac{22}{13}\right]\right), \left(3, \left[\frac{33}{13}\right]\right),$$
$$\left(4, \left[\frac{44}{13}\right]\right), \left(5, \left[\frac{55}{13}\right]\right), \left(6, \left[\frac{66}{13}\right]\right)$$

on a graph.

48 If we refer to points for which both coordinates are integers as *lattice points*, determine whether there are lattice points on any of the lines

$y = \frac{11}{13}x$, $y = \frac{11}{13}x + \frac{1}{2}$, $y = \frac{11}{13}x - \frac{1}{2}$, for $0 < x < 13$?

49 How many lattice points are there between the lines $y = \frac{11}{13}x - \frac{1}{2}$ and $y = \frac{11}{13}x + \frac{1}{2}$ for $1 \leqslant x \leqslant 6$?

50 Use q 47, sharpened if necessary by the calculations of q 44, to determine how many lattice points there are between the lines $y = \frac{11}{13}x$ and $y = \frac{11}{13}x + \frac{1}{2}$ for $1 \leqslant x \leqslant 6$. How does this number relate to the number of minus signs in the list of numerically least residues congruent to 11, 22, 33, 44, 55, 66 (modulo 13)? Does this number enable you to determine whether 11 is a quadratic residue modulo 13?

51 Express $\frac{13}{11} - [\frac{13}{11}]$, $\frac{26}{11} - [\frac{26}{11}]$, $\frac{39}{11} - [\frac{39}{11}]$, $\frac{52}{11} - [\frac{52}{11}]$, $\frac{65}{11} - [\frac{65}{11}]$ as simple fractions. How many of these numbers exceed $\frac{1}{2}$? If $x/11 - [x/11]$ exceeds $\frac{1}{2}$, what can be said about the numerically least residue congruent to x (mod 11)? Use these results to determine whether 13 is a quadratic residue modulo 11. Check your result with table 3.2.

52 Using paper similar to that used in q 47, but a clean sheet, mark the points $(\frac{13}{11}, 1)$, $(\frac{26}{11}, 2)$, $(\frac{39}{11}, 3)$, $(\frac{52}{11}, 4)$, $(\frac{65}{11}, 5)$ and the points $([\frac{13}{11}], 1)$, $([\frac{26}{11}], 2)$, $([\frac{39}{11}], 3)$, $([\frac{52}{11}], 4)$, $([\frac{65}{11}], 5)$.

53 Are there any lattice points on the lines

$x = \frac{13}{11}y$, $x = \frac{13}{11}y - \frac{1}{2}$, $x = \frac{13}{11}y + \frac{1}{2}$, for $0 < y < 11$?

54 How many lattice points are there between the lines

$x = \frac{13}{11}y - \frac{1}{2}$ and $x = \frac{13}{11}y + \frac{1}{2}$ for $1 \leqslant y \leqslant 5$?

⟦90⟧

55 Use q 52, sharpened if necessary by the calculations of q 51, to determine how many lattice points there are between the lines $x = \frac{13}{11}y$ and $x = \frac{13}{11}y + \frac{1}{2}$ for $1 \leq y \leq 5$. How does this number enable you to determine whether 13 is a quadratic residue modulo 11?

56 Make a sketch graph of the rectangle $1 \leq x \leq 6$, $1 \leq y \leq 5$, and mark roughly where the lines $y = \frac{11}{13}x + \frac{1}{2}$, $y = \frac{11}{13}x$ and $x = \frac{13}{11}y + \frac{1}{2}$ lie in relation to this rectangle. How many lattice points are there inside and on the perimeter of the rectangle? Under the half-turn with centre $(3\frac{1}{2}, 3)$, that is $(x, y) \to (-x+7, -y+6)$, what is the image of the rectangle, and what is the image of the line $y = \frac{11}{13}x + \frac{1}{2}$?

57 Prove that the total number of lattice points in the rectangle $1 \leq x \leq 6$, $1 \leq y \leq 5$ is even or odd according as the number of lattice points within the rectangle and between the lines $y = \frac{11}{13}x + \frac{1}{2}$ and $x = \frac{13}{11}y + \frac{1}{2}$ is even or odd.

58 If p and q are distinct odd prime numbers and the number of lattice points between $y = (p/q)x$ and $y = (p/q)x + \frac{1}{2}$, where $1 \leq x \leq \frac{1}{2}(q-1)$, is l, use Gauss' lemma to explain why $\left(\frac{p}{q}\right) = (-1)^l$.

59 If p and q are odd prime numbers and the number of lattice points between $x = (q/p)y$ and $x = (q/p)y + \frac{1}{2}$, where $1 \leq y \leq \frac{1}{2}(p-1)$, is m, explain why $\left(\frac{q}{p}\right) = (-1)^m$. Deduce that $\left(\frac{p}{q}\right)\left(\frac{q}{p}\right) = (-1)^{l+m}$.

60 Verify that the half-turn about the point $(\frac{1}{4}(q+1), \frac{1}{4}(p+1))$, namely, $(x, y) \to (-x + \frac{1}{2}(q+1), -y + \frac{1}{2}(p+1))$ maps the rectangle $1 \leq x \leq \frac{1}{2}(q-1)$, $1 \leq y \leq \frac{1}{2}(p-1)$ onto itself and interchanges the lines $y = (p/q)x + \frac{1}{2}$ and $x = (q/p)y + \frac{1}{2}$. Deduce that $l + m$, as defined in q 58 and q 59, is even or odd according as $\frac{1}{4}(p-1)(q-1)$ is even or odd, so that $(-1)^{l+m} = (-1)^{\frac{1}{4}(p-1)(q-1)}$.

61 Using table 3.2, find $(\frac{3}{5})$ and $(\frac{5}{3})$, $(\frac{3}{7})$ and $(\frac{7}{3})$, $(\frac{5}{7})$ and $(\frac{7}{5})$, $(\frac{3}{11})$ and $(\frac{11}{3})$, $(\frac{5}{11})$ and $(\frac{11}{5})$, $(\frac{7}{11})$ and $(\frac{11}{7})$, $(\frac{3}{13})$ and $(\frac{13}{3})$, $(\frac{5}{13})$ and $(\frac{13}{5})$, $(\frac{7}{13})$ and $(\frac{13}{7})$.

Isolate those cases where $\left(\frac{p}{q}\right) \neq \left(\frac{q}{p}\right)$.

62 Use q 60, to determine precisely when $\left(\frac{p}{q}\right) \neq \left(\frac{q}{p}\right)$.

(*Law of quadratic reciprocity*)

[[91]]

63 Justify each step of the following argument.

$$\begin{aligned}\left(\frac{350}{19}\right) &= \left(\frac{2\cdot 5\cdot 5\cdot 7}{19}\right)\\ &= \left(\frac{2}{19}\right)\left(\frac{5}{19}\right)\left(\frac{5}{19}\right)\left(\frac{7}{19}\right)\\ &= \left(\frac{2}{19}\right)\left(\frac{7}{19}\right)\\ &= -\left(\frac{7}{19}\right)\\ &= \left(\frac{19}{7}\right)\\ &= \left(\frac{5}{7}\right)\\ &= \left(\frac{7}{5}\right)\\ &= \left(\frac{2}{5}\right)\\ &= -1.\end{aligned}$$

64 Find $\left(\frac{3}{p}\right)$ classifying the answers according to whether $p \equiv 1, 5, 7, 11 \pmod{12}$.

65 Find $\left(\frac{5}{p}\right)$ classifying the answers according to whether $p \equiv 1, 3, 7, 9 \pmod{10}$.

Notes and answers

For concurrent reading see bibliography: Shanks (1978), Davenport (1968), Bolker (1970).

1 $0^2=0$, $1^2=6^2=1$, $3^2=4^2=2$, $2^2=5^2=4$. The use of the equals sign, '=', rather than the congruence sign, '≡', indicates that we are considering the elements of $\mathbf{Z}_7$ to be the seven residue classes modulo 7.

2 No solution: (iv), (vi), (vii), (xii).
One solution: (i), (viii), (ix).
Two solutions: (ii), (iii), (v), (x), (xi).

3

1	2	3	4	5	6
1	4	2	2	4	1

Subgroup

1	4	2
4	2	1
2	1	4

The perfect squares in $\mathbf{M}_n$ are called quadratic residues modulo n.

4 All subgroups

mod 5

1	4
4	1

mod 12

1

mod 13

1	4	9	3	12	10
4	3	10	12	9	1
9	10	3	1	4	12
3	12	1	9	10	4
12	9	4	10	1	3
10	1	12	4	3	9

mod 11

1	4	9	5	3
4	5	3	9	1
9	3	4	1	5
5	9	1	3	4
3	1	5	4	9

Two to one: 5, 11, 13.
Four to one: 12.

5 1, 2, 3, 5, 6, 8. $\frac{1}{2}(p-1)$.

6 $x^2 \equiv y^2 \Rightarrow x^2 - y^2 \equiv 0 \Rightarrow (x+y)(x-y) \equiv 0 \pmod p$
$\Rightarrow p \mid x+y$ or $p \mid x-y$.

7 $x^2 \equiv 1 \pmod p \Rightarrow p \mid x+1$ or $p \mid x-1 \Rightarrow x \equiv \pm 1 \pmod p$.
Exactly two, from q 6.

8 5, 7, 11.

9 $(\text{q.r.})^3 \equiv 1 \pmod 7$, $(\text{q.n-r.})^3 \equiv -1 \pmod 7$.

10 There are exactly six roots in $\mathbf{M}_7$.
$(x^2)^3 \equiv 1 \pmod 7$ for all $x \in \mathbf{M}_7$, so the three quadratic residues are solutions of $x^3 \equiv 1$. $x^6 - 1 = (x^3-1)(x^3+1)$. By Lagrange's theorem $x^3 \equiv -1$ must have exactly three roots, so these are the quadratic non-residues.

11 $x^{10} - 1 \equiv 0 \pmod{11}$ has ten solutions by Fermat. The five quadratic residues satisfy $x^5 \equiv 1$. $x^{10} - 1 = (x^5-1)(x^5+1)$. So $x^5 \equiv -1$ has exactly five solutions, which must be the five quadratic non-residues.

12 $x^{p-1}-1\equiv 0 \pmod p$ has $p-1$ roots by Fermat. The $\frac{1}{2}(p-1)$ quadratic residues, from q 6, satisfy $x^{\frac{1}{2}(p-1)}\equiv 1 \pmod p$, and since $x^{p-1}-1=(x^{\frac{1}{2}(p-1)}-1)(x^{\frac{1}{2}(p-1)}+1)$, $x^{\frac{1}{2}(p-1)}\equiv -1 \pmod p$ has exactly $\frac{1}{2}(p-1)$ roots by Lagrange, which must be the $\frac{1}{2}(p-1)$ quadratic non-residues.

13 $x^k y^k=(xy)^k$ gives closure. $1^k=1$.
If $ax\equiv 1$, then $(ax)^k\equiv 1$, so $a^k x^k\equiv 1$ and the inverse of each kth power is a kth power.
This establishes in particular that the quadratic residues in $\mathbf{M}_n$ form a subgroup.

14 $\{1, -1\}$.

15 If $a\equiv 0 \pmod p$, then $ab\equiv 0 \pmod p$ and

$$\left(\frac{ab}{p}\right)=0=0\cdot\left(\frac{b}{p}\right)=\left(\frac{a}{p}\right)\left(\frac{b}{p}\right).$$

If $a\not\equiv 0 \pmod p$, then $\left(\frac{a}{p}\right)\equiv a^{\frac{1}{2}(p-1)} \pmod p$ by q 12, so if $a, b\not\equiv 0 \pmod p$,

$$\left(\frac{ab}{p}\right)\equiv (ab)^{\frac{1}{2}(p-1)}=a^{\frac{1}{2}(p-1)}b^{\frac{1}{2}(p-1)}\equiv\left(\frac{a}{p}\right)\left(\frac{b}{p}\right).$$

For the definition of the Legendre symbol, p is an odd prime number. If p is not necessarily prime, a similar symbol with some of the same properties is called the *Jacobi symbol.*

16 -1 is a quadratic residue for 5, 13, $17\equiv 1 \pmod 4$.
-1 is a quadratic non-residue for 3, 7, 11, $19\equiv 3 \pmod 4$.

17 Two. There are no other elements of order 2, from q 7.

18 (i) $\frac{1}{2}(p-1)$ must be even, so $4|p-1$ and $p\equiv 1 \pmod 4$.
(ii) $\frac{1}{2}(p-1)$ must be odd, so $p-1=2(2k+1)$, $p\equiv 3 \pmod 4$.

19 The subgroup of quadratic residues has order $\frac{1}{2}(p-1)$. This is even if $p\equiv 1 \pmod 4$ and odd if $p\equiv 3 \pmod 4$.

20 The order of an element divides the order of the group.

21 Unless an element has order 1 or 2, it has a distinct inverse. So a group contains an even number of elements with order different from 1 and 2. If the whole group is of even order, then the elements of order 1 and 2 are even in number. But 1 is of order 1, so there is at least one element of order 2.

22 If $p=4k+1$, the group of quadratic residues has order $2k$. Since this is an even number, the group contains an element of order 2. The only such element is -1, so -1 is a quadratic residue and $a^2\equiv -1 \pmod p$ for some a. The element a has order 4 in $\mathbf{M}_p$.

Or, if b is a quadratic non-residue, $b^{2k} \equiv -1 \pmod p$ by q 12, so b^k does not have order 1 or 2. However, $b^{4k} \equiv 1$ by Fermat's theorem, so b^k has order 4.

23 5, 5, 17, 13, 37, 5, 5 and 13, 41, each congruent to 1 (mod 4).

24 If $p|n^2+1$, $n^2 \equiv -1 \pmod p$, so -1 is a quadratic residue and $p \equiv 1 \pmod 4$ from q 18 and q 12.

25 Using the conventions of q 1.64, $p_1^2 p_2^2 p_3^2 \ldots p_n^2 + 1$ is not divisible by any prime $\leqslant p_n$. Each of its prime factors is of the form $4k+1$ by q 24 and is greater than p_n, so the number of primes of this form is infinite.

26 7, 13, $3 \cdot 7$, 31, 43, $3 \cdot 19$, 73, $7 \cdot 13$.

All but 3 are of the form $3k+1$.

27 If $p|n^2+n+1$, then $p|n^3-1=(n-1)(n^2+n+1)$, so n is of order 1 or 3 in $\mathbf{M}_p$. If n is of order 1, $p=3$. If n is of order 3, $\mathbf{M}_p$ contains an element of order 3, and $3|p-1$, so $p \equiv 1 \pmod 3$.

28 With the conventions of q 1.64, let $m = p_1 p_2 \ldots p_n$, then $m(m+1)+1$ has no prime factor $\leqslant p_n$. Each of its prime factors has the form $3k+1$ by q 27 and is greater than p_n, so the number of primes of this form is infinite.

29 By Wilson's theorem $1 \cdot 2 \ldots 4k \equiv -1 \pmod p$. But $2k+1 \equiv -2k, \ldots, -2 \equiv 4k-1, -1 \equiv 4k$, so $1 \cdot 2 \ldots 2k \cdot -2k \ldots -2 \cdot -1 \equiv -1 \pmod p$.
There are $2k$ minus signs here, so

$(1 \cdot 2 \ldots 2k)^2 \equiv -1 \pmod p$.

This establishes that $1 \cdot 2 \ldots 2k$ is an element of order 4 in $\mathbf{M}_p$.

What needs to be remembered from the last thirteen questions is that -1 is a quadratic residue modulo an odd prime congruent to 1 (mod 4), and a quadratic non-residue modulo an odd prime congruent to 3 (mod 4). This result will be used in chapter 6.

30 Each row contains an even number, $p-1$, of elements. $p-1 \equiv -1$, $p-2 \equiv -2, \ldots, p-\frac{1}{2}(p-1) \equiv -\frac{1}{2}(p-1)$, so the second half recapitulates the absolute values of the first half. In a later row, each entry, ab, in the first half is matched by $a(-b) = -(ab)$ in the second half. $x \to ax$ is a permutation of $\mathbf{M}_p$, so every absolute value must appear exactly twice, and thus the first half contains all possible absolute values, once only.

31

	Even	Odd
Mod 3	1	−1
Mod 5	1, −1	2, −2
Mod 7	1, 2, −3	3, −2, −1
Mod 11	1, 3, 4, 5, −2	2, −5, −4, −3, −1
Mod 13	1, 3, 4, −4, −3, −1	2, 5, 6, −2, −5, −6
	Quadratic residues	Quadratic non-residues

32 2, 4, 6, 8, −7, −5, −3, −1.
So $2^8(1\cdot2\cdot3\cdot4\cdot5\cdot6\cdot7\cdot8)\equiv(-1)^4(1\cdot2\cdot3\cdot4\cdot5\cdot6\cdot7\cdot8)$ (mod 17).
Since $2^8\equiv(-1)^4\equiv1$ (mod 17), 2 is a quadratic residue.

33 $3^8(1\cdot2\cdot3\cdot4\cdot5\cdot6\cdot7\cdot8)\equiv(-1)^3(1\cdot2\cdot3\cdot4\cdot5\cdot6\cdot7\cdot8)$ (mod 17).
So $3^8\equiv(-1)^3\equiv-1$ (mod 17) and 3 is a quadratic non-residue.

34 $2\cdot1\equiv2$, $2\cdot2\equiv4$, $2\cdot3\equiv6$, $2\cdot4\equiv8$, $2\cdot5\equiv-9$, $2\cdot6\equiv-7$, $2\cdot7\equiv-5$, $2\cdot8\equiv-3$, $2\cdot9\equiv-1$ (mod 19).
So $2^9(1\cdot2\cdot3\cdot4\cdot5\cdot6\cdot7\cdot8\cdot9)\equiv(-1)^5(1\cdot2\cdot3\cdot4\cdot5\cdot6\cdot7\cdot8\cdot9)$ (mod 19). Since $2^9\equiv-1$ (mod 19), 2 is a quadratic non-residue.

35 Distinct absolute values were established in q 30. If the number of minus signs among the numerically least residues is m,
$a^8(1\cdot2\cdot3\cdot4\cdot5\cdot6\cdot7\cdot8)\equiv(-1)^m(1\cdot2\cdot3\cdot4\cdot5\cdot6\cdot7\cdot8)$ (mod 17),
so $a^8\equiv(-1)^m$ (mod 17). If m is even, $a^8\equiv1$ and a is a quadratic residue. If m is odd, $a^8\equiv-1$ and a is a quadratic non-residue.

36 Distinct absolute values as in q 30.
If there are m minus signs among the numerically least residues, $2^{\frac{1}{2}(p-1)}\equiv(-1)^m \pmod p$ and 2 is a quadratic residue or not according as m is even or odd.

37 If $p=8k+1$, numerically least residues lie between $\pm4k$.
$2\cdot1, 2\cdot2, \ldots, 2\cdot4k\equiv2, 4, \ldots, 4k, -4k+1, -4k+3, \ldots, -3, -1$.
The numerically least residues use exactly $2k$ minus signs, so $2^{4k}\equiv(-1)^{2k}\equiv1 \pmod p$ and 2 is a quadratic residue.

38 If $p=8k+3$, numerically least residues lie between $\pm(4k+1)$.
$2\cdot1, \ldots, 2(4k+1)\equiv2, 4, \ldots, 4k, -4k-1, -4k+1, \ldots, -3, -1$.
The numerically least residues use exactly $2k+1$ minus signs, so $2^{4k+1}\equiv(-1)^{2k+1}\equiv-1 \pmod p$ and 2 is a quadratic non-residue.

39 If $p=8k+5$, numerically least residues lie between $\pm(4k+2)$.
$2\cdot1, \ldots, 2(4k+2)\equiv2, 4, \ldots, 4k+2, -4k-1, -4k+1, \ldots, -3, -1$.
The numerically least residues use exactly $2k+1$ minus signs, so $2^{4k+2}\equiv(-1)^{2k+1}\equiv-1 \pmod p$ and 2 is a quadratic non-residue.

40 If $p=8k+7$, numerically least residues lie between $\pm(4k+3)$. $2\cdot1, \ldots, 2(4k+3)\equiv2, 4, \ldots, 4k+2, -4k-3, -4k-1, -4k+1, \ldots, -3, -1$.
The numerically least residues use exactly $2k+2$ minus signs, so $2^{4k+3}\equiv(-1)^{2k+2}\equiv1 \pmod p$ and 2 is a quadratic residue.

41 If $p=8k+1$, $(p-1)(p+1)=8k(8k+2)$, so $\frac{1}{8}(p^2-1)=2k(4k+1)$, even.
If $p=8k+3$, $(p-1)(p+1)=(8k+2)(8k+4)$, $\frac{1}{8}(p^2-1)=(4k+1)(2k+1)$, odd.
If $p=8k+5$, $(p-1)(p+1)=(8k+4)(8k+6)$, $\frac{1}{8}(p^2-1)=(2k+1)(4k+3)$, odd.
If $p=8k+7$, $(p-1)(p+1)=(8k+6)(8k+8)$, $\frac{1}{8}(p^2-1)=2(4k+3)(k+1)$, even.
Result then follows from, q 37, q 38, q 39 and q 40.

42

	$p=$	$8k+1$	$8k+3$	$8k+5$	$8k+7$
	$\left(\frac{-1}{p}\right)=$	1	−1	1	−1
	$\left(\frac{2}{p}\right)=$	1	−1	−1	1
so	$\left(\frac{-2}{p}\right)=$	1	1	−1	−1

43 Distinct absolute values by q 30, so $a^{\frac{1}{2}(p-1)}[1\cdot 2\cdot 3\ldots\frac{1}{2}(p-1)]\equiv(-1)^l[1\cdot 2\ldots\frac{1}{2}(p-1)] \pmod p$ and $a^{\frac{1}{2}(p-1)}\equiv(-1)^l \pmod p$, so a is a quadratic residue if l is even, and a quadratic non-residue if l is odd.

44 $\frac{11}{13}, \frac{9}{13}, \frac{7}{13}, \frac{5}{13}, \frac{3}{13}, \frac{1}{13}$. The first three.

45 If $x\equiv a \pmod{13}$ and $-6\leqslant a\leqslant -1$, then $x-a\equiv 0 \pmod{13}$ and $(x-a)/13$ is an integer $>x/13$.

$$\text{So } \frac{x}{13}-\left[\frac{x}{13}\right]=\frac{x}{13}-\left(\frac{x-a}{13}-1\right)=\frac{a}{13}+1>\frac{1}{2}.$$

If $x\equiv a \pmod{13}$ and $1\leqslant a\leqslant 6$, then $(x-a)/13$ is an integer $<x/13$, so

$$\frac{x}{13}-\left[\frac{x}{13}\right]=\frac{x}{13}-\frac{x-a}{13}=\frac{a}{13}<\frac{1}{2}.$$

46 Since $\frac{11}{13}, \frac{9}{13}, \frac{7}{13}>\frac{1}{2}$, from q 44 and q 45, the numerically least residues congruent to 11, 22 and 33 are negative, and the numerically least residues congruent to 44, 55 and 66 are positive. Thus from q 43, $11^6\equiv(-1)^3 \pmod{13}$, and 11 is a quadratic non-residue modulo 13.

48 $x=1, 2, \ldots, 12$ in $y=\frac{11}{13}x$, gives $y=\frac{11}{13}, \frac{22}{13}, \ldots, \frac{132}{13}$. None of these values for y are integers or integers $\pm\frac{1}{2}$, since 13 does not divide $11x$.

49 Six, because the lines are 1 unit apart in the y-direction.

50 Three. For given x, there is a lattice point between $y=\frac{11}{13}x$ and $y=\frac{11}{13}x+\frac{1}{2}$ or between $y=\frac{11}{13}x$ and $y=\frac{11}{13}x-\frac{1}{2}$ but not both. There is a lattice point between the second pair when $\frac{11}{13}x-[\frac{11}{13}x]<\frac{1}{2}$, so there is a lattice point between the first pair when $\frac{11}{13}x-[\frac{11}{13}x]>\frac{1}{2}$, and this happens when the numerically least residue congruent to $11x \pmod{13}$ is negative.

51 $\frac{2}{11}, \frac{4}{11}, \frac{6}{11}, \frac{8}{11}, \frac{10}{11}$. The last three are $>\frac{1}{2}$.

$$1>\frac{x}{11}-\left[\frac{x}{11}\right]>\frac{1}{2} \Leftrightarrow 0>\frac{x}{11}-\left[\frac{x}{11}+1\right]>-\frac{1}{2}$$

$$\Leftrightarrow 0>x-11\left[\frac{x}{11}+1\right]>-\frac{11}{2}.$$

So the numerically least residue congruent to x (mod 11) is negative. Thus the numerically least residues congruent to 39, 52 and 65 (mod 11) are negative, and by Gauss' lemma, 13 is not a quadratic residue modulo 11.

53 No.

54 Five, because the lines are 1 unit apart in the x-direction.

55 There are three, by an argument similar to that of q 50, and these occur at those values of y such that the numerically least residues congruent to $13y$ (mod 11) are negative, by q 51. So 13 is a quadratic non-residue mod 11.

56 There are thirty lattice points in the rectangle (see fig. 4.1). $(3\frac{1}{2}, 3)$ is the centre of the rectangle. The equation $y=\frac{11}{13}x+\frac{1}{2}$ is equivalent to $(-x+7)=\frac{13}{11}(-y+6)+\frac{1}{2}$. So (x, y) lies on the line $y=\frac{11}{13}x+\frac{1}{2}$ if and only if $(-x+7, -y+6)$ lies on the line $x=\frac{13}{11}y+\frac{1}{2}$, and the rectangle is mapped to itself and the lines interchanged by the half-turn.

Fig. 4.1

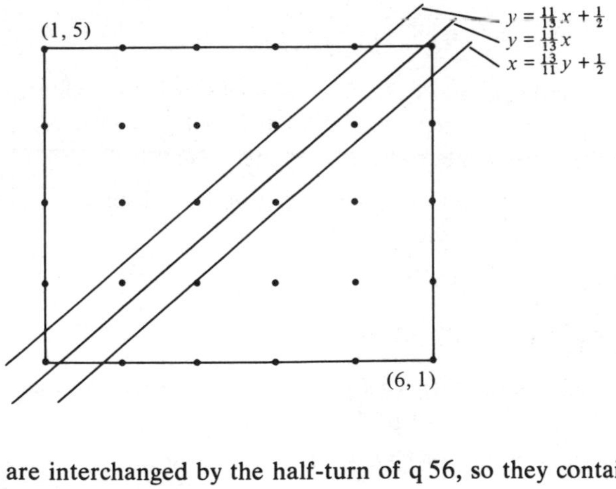

57 A and B are interchanged by the half-turn of q 56, so they contain the same number of lattice points.
Number of lattice points in A and $B \equiv 0 \pmod 2$.

Number of lattice points in A and B and $C \equiv$ Number of lattice points in C (mod 2). In this case, we have $30 \equiv 6 \pmod 2$. (See fig. 4.2.)

Fig. 4.2

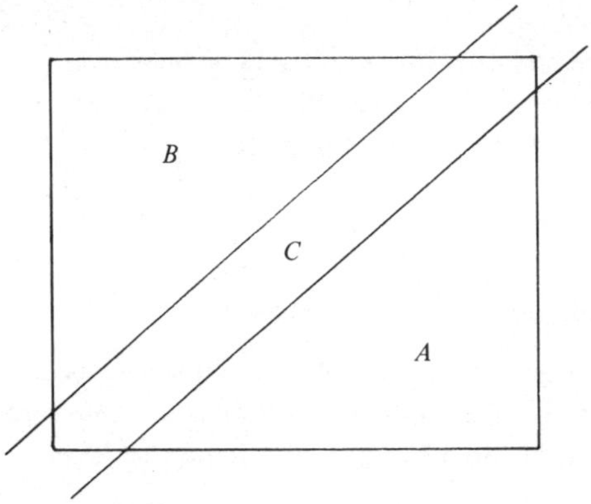

58 For a given integer x, there is a lattice point between $y=(p/q)x$ and $y=(p/q)x+\frac{1}{2}$ when $(p/q)x-[(p/q)x]>\frac{1}{2}$. There are no lattice points on these lines for $1 \leqslant x \leqslant q-1$, since p and q are distinct odd primes. Generalising the argument of q 45, $(p/q)x-[(p/q)x]>\frac{1}{2}$ precisely when the numerically least residue congruent to $px \pmod q$ is negative. So if there are l lattice points in the region, $\left(\frac{p}{q}\right)=(-1)^l$ by Gauss' lemma.

59 $\left(\frac{q}{p}\right)=(-1)^m$ as in q 58, so $\left(\frac{p}{q}\right)\left(\frac{q}{p}\right)=(-1)^{l+m}$.

60 From the diagram of q 57, C contains $l+m$ lattice points, the whole rectangle contains $\frac{1}{4}(p-1)(q-1)$ lattice points, and since A and B contain the same number of lattice points $l+m \equiv \frac{1}{4}(p-1)(q-1) \pmod 2$.

61 -1 and -1, -1 and 1, -1 and -1, 1 and -1, 1 and 1, -1 and 1, 1 and 1, -1 and -1, -1 and -1.

$(\frac{3}{7}) \neq (\frac{7}{3})$, $(\frac{3}{11}) \neq (\frac{11}{3})$, $(\frac{7}{11}) \neq (\frac{11}{7})$.

62 $\left(\frac{p}{q}\right)=\left(\frac{q}{p}\right)$ unless $\frac{1}{4}(p-1)(q-1)$ is odd.

If $p=4k+1$, then $\frac{1}{4}(p-1)(q-1)=k(q-1)$, which is even.

If $p=4k+3$ and $q=4h+3$, then

$\frac{1}{4}(p-1)(q-1)=(2k+1)(2h+1)$, which is odd.

So $\left(\frac{p}{q}\right)=\left(\frac{q}{p}\right)$ unless $p \equiv q \equiv 3 \pmod 4$.

63 (i) Factors of 350, (ii) from q 15, (iii) $\left(\frac{a^2}{19}\right)=1$, (iv) $\left(\frac{2}{19}\right)=-1$ from q 38, (v) law of quadratic reciprocity, (vi) $19\equiv 5 \pmod 7$, (vii) law of quadratic reciprocity, (viii) $7\equiv 2 \pmod 5$, (ix) $\left(\frac{2}{5}\right)=-1$ from q 39.

64 $\left(\frac{3}{12k+1}\right)=\left(\frac{12k+1}{3}\right)=\left(\frac{1}{3}\right)=1.$

$\left(\frac{3}{12k+5}\right)=\left(\frac{12k+5}{3}\right)=\left(\frac{2}{3}\right)=-1.$

$\left(\frac{3}{12k+7}\right)=-\left(\frac{12k+7}{3}\right)=-\left(\frac{1}{3}\right)=-1.$

$\left(\frac{3}{12k+11}\right)=-\left(\frac{12k+11}{3}\right)=-\left(\frac{2}{3}\right)=1.$

65 $\left(\frac{5}{10k+1}\right)=\left(\frac{10k+1}{5}\right)=\left(\frac{1}{5}\right)=1.$

$\left(\frac{5}{10k+3}\right)=\left(\frac{10k+3}{5}\right)=\left(\frac{3}{5}\right)=\left(\frac{5}{3}\right)=\left(\frac{2}{3}\right)=-1.$

$\left(\frac{5}{10k+7}\right)=\left(\frac{10k+7}{5}\right)=\left(\frac{2}{5}\right)=-1.$

$\left(\frac{5}{10k+9}\right)=\left(\frac{10k+9}{5}\right)=\left(\frac{4}{5}\right)=\left(\frac{2^2}{5}\right)=1.$

Historical note

Although Euler and Lagrange knew the law of quadratic reciprocity, and Legendre used it in his investigations of quadratic forms, it was Gauss who provided the first proof in 1801. He constructed several other proofs in the course of his life, and some of these are given in Mathews (no date).

Table 5.1

0																				
1	2																			
4	5	8																		
9	10	13	18																	
16	17	20	25	32																
25	26	29	34	41	50															
36	37	40	45	52	61	72														
49	50	53	58	65	74	85	98													
64	65	68	73	80	89	100	113	128												
81	82	85	90	97	106	117	130	145	162											
100	101	104	109	116	125	136	149	164	181	200										
121	122	125	130	137	146	157	170	185	202	221	242									
144	145	148	153	160	169	180	193	208	225	244	265	288								
169	170	173	178	185	194	205	218	233	250	269	290	313	338							
196	197	200	205	212	221	232	245	260	277	296	317	340	365	392						
225	226	229	234	241	250	261	274	289	306	325	346	369	394	421	450					
256	257	260	265	272	281	292	305	320	337	356	377	400	425	452	481	512				
289	290	293	298	305	314	325	338	353	370	389	410	433	458	485	514	545	578			
324	325	328	333	340	349	360	373	388	405	424	445	468	493	520	549	580	613	648		
361	362	365	370	377	386	397	410	425	442	461	482	505	530	557	586	617	650	685	722	
400	401	404	409	416	425	436	449	464	481	500	521	544	569	596	625	656	689	724	761	800
441	442	445	450	457	466	477	490	505	522	541	562	585	610	637	666	697	730	765	802	841
484	485	488	493	500	509	520	533	548	565	584	605	628	653	680	709	740	773	808	845	884
529	530	533	538	545	554	565	578	593	610	629	650	673	698	725	754	785	818	853	890	929
576	577	580	585	592	601	612	625	640	657	676	697	720	745	772	801	832	865	900	937	976
625	626	629	634	641	650	661	674	689	706	725	746	769	794	821	850	881	914	949	986	1025
676	677	680	685	692	701	712	725	740	757	776	797	820	845	872	901	932	965	1000	1037	1076
729	730	733	738	745	754	765	778	793	810	829	850	873	898	925	954	985	1018	1053	1090	1129
784	785	788	793	800	809	820	833	848	865	884	905	928	953	980	1009	1040	1073	1108	1145	1184
841	842	845	850	857	866	877	890	905	922	941	962	985	1010	1037	1066	1097	1130	1165	1202	1241
900	901	904	909	916	925	936	949	964	981	1000	1021	1044	1069	1096	1125	1156	1189	1224	1261	1300
961	962	965	970	977	986	997	1010	1025	1042	1061	1082	1105	1130	1157	1186	1217	1250	1285	1322	1361
1024	1025	1028	1033	1040	1049	1060	1073	1088	1105	1124	1145	1168	1193	1220	1249	1280	1313	1348	1385	1424
1089	1090	1093	1098	1105	1114	1125	1138	1153	1170	1189	1210	1233	1258	1285	1314	1345	1378	1413	1450	1489
1156	1157	1160	1165	1172	1181	1192	1205	1220	1237	1256	1277	1300	1325	1352	1381	1412	1445	1480	1517	1556
1225	1226	1229	1234	1241	1250	1261	1274	1289	1306	1325	1346	1369	1394	1421	1450	1481	1514	1549	1586	1625
1296	1297	1300	1305	1312	1321	1332	1345	1360	1377	1396	1417	1440	1465	1492	1521	1552	1585	1620	1657	1696
1369	1370	1373	1378	1385	1394	1405	1418	1433	1450	1469	1490	1513	1538	1565	1594	1625	1658	1693	1730	1769
1444	1445	1448	1453	1460	1469	1480	1493	1508	1525	1544	1565	1588	1613	1640	1669	1700	1733	1768	1805	1844
1521	1522	1525	1530	1537	1546	1557	1570	1585	1602	1621	1642	1665	1690	1717	1746	1777	1810	1845	1882	1921
1600	1601	1604	1609	1616	1625	1636	1649	1664	1681	1700	1721	1744	1769	1796	1825	1856	1889	1924	1961	2000
1681	1682	1685	1690	1697	1706	1717	1730	1745	1762	1781	1802	1825	1850	1877	1906	1937	1970	2005	2042	2081
1764	1765	1768	1773	1780	1789	1800	1813	1828	1845	1864	1885	1908	1933	1960	1989	2020	2053	2088	2125	2164
1849	1850	1853	1858	1865	1874	1885	1898	1913	1930	1949	1970	1993	2018	2045	2074	2105	2138	2173	2210	2249
1936	1937	1940	1945	1952	1961	1972	1985	2000	2017	2036	2057	2080	2105	2132	2161	2192	2225	2260	2297	2336
2025	2026	2029	2034	2041	2050	2061	2074	2089	2106	2125	2146	2169	2194	2221	2250	2281	2314	2349	2386	2425
2116	2117	2120	2125	2132	2141	2152	2165	2180	2197	2216	2237	2260	2285	2312	2341	2372	2405	2440	2477	2516
2209	2210	2213	2218	2225	2234	2245	2258	2273	2290	2309	2330	2353	2378	2405	2434	2465	2498	2533	2570	2609
2304	2305	2308	2313	2320	2329	2340	2353	2368	2385	2404	2425	2448	2473	2500	2529	2560	2593	2628	2665	2704
2401	2402	2405	2410	2417	2426	2437	2450	2465	2482	2501	2522	2545	2570	2597	2626	2657	2690	2725	2762	2801
2500	2501	2504	2509	2516	2525	2536	2549	2564	2581	2600	2621	2644	2669	2696	2725	2756	2789	2824	2861	2900

5

The equation $x^n + y^n = z^n$, for $n = 2, 3, 4$

The equation $x^2 + y^2 = z^2$

1 Table 5.1 is a table of sums of squares. Using the left hand column of squares as a checklist, determine the location of all the square numbers in the table. The only square which appears that is greater than 2500 is 2704.

2 What are the square numbers which appear in table 5.1 along a line joining 0 to $4^2+3^2=5^2$? Given the first of these could you have predicted the others?

3 What are the square numbers which appear along a line joining 0 to $12^2+5^2=13^2$? What is the similarity between them?

4 Use the fact that $24^2+7^2=25^2$ and $15^2+8^2=17^2$ to predict two more square numbers in table 5.1.

5 If x, y and z are positive integers such that $x^2+y^2=z^2$, then (x, y, z) is called a *Pythagorean triple.* If, moreover, $\gcd(x, y, z)=1$, then (x, y, z) is called a *primitive* Pythagorean triple. Identify the seven distinct primitive Pythagorean triples which may be deduced from table 5.1, ignoring the possibility that $(x, y, z) \neq (y, x, z)$.

6 If (x, y, z) is a Pythagorean triple, is it possible that
(i) all three of x, y and z are even,
(ii) just two of x, y and z are even,
(iii) just one of x, y and z is even,
(iv) neither x, nor y nor z is even?
Either illustrate the possibility, or prove the impossibility.

7 If (x, y, z) is a Pythagorean triple, is it possible that
(i) all three of x, y and z are divisible by 3,
(ii) just two of x, y and z are divisible by 3,
(iii) just one of x, y and z is divisible by 3,
(iv) neither x, nor y nor z is divisible by 3?
Either illustrate the possibility, or prove the impossibility. For (iv), make a table of sums of squares in $\mathbf{Z}_3$.

8 If (x, y, z) is a Pythagorean triple, is it possible that
(i) all three of x, y and z are divisible by 5,
(ii) just two of x, y and z are divisible by 5,
(iii) just one of x, y and z is divisible by 5,
(iv) neither x, nor y nor z is divisible by 5?
Either illustrate the possibility, or prove the impossibility. For (iv), make a table of sums of squares in $\mathbf{Z}_5$.

9 If (x, y, z) is a Pythagorean triple, prove that z is even only when x and y are even, by making a table of sums of squares in $\mathbf{Z}_4$. If x is even, but y and z are odd, prove that x has a factor 4 by considering $x^2 = z^2 - y^2$.

10 If (x, y, z) is a Pythagorean triple and $\sin\theta = x/z$, $\tan\theta = x/y$ (see fig. 5.1), use the fact that $x : y : z = 2\tan\frac{1}{2}\theta : 1 - \tan^2\frac{1}{2}\theta : 1 + \tan^2\frac{1}{2}\theta$ to verify that $\tan\frac{1}{2}\theta = x/(y+z)$.
If $\tan\frac{1}{2}\theta = q/p$, where p and q are integers such that $\gcd(p, q) = 1$, deduce that $x : y : z = 2pq : p^2 - q^2 : p^2 + q^2$.

Fig. 5.1

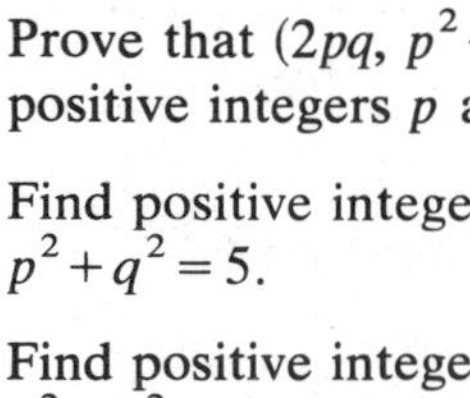

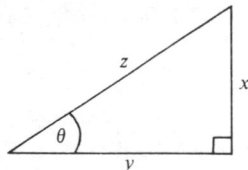

11 Prove that $(2pq, p^2 - q^2, p^2 + q^2)$ is a Pythagorean triple for any positive integers p and q, with $p > q$.

12 Find positive integers p and q such that $2pq = 4$, $p^2 - q^2 = 3$ and $p^2 + q^2 = 5$.

13 Find positive integers p and q such that $2pq = 12$, $p^2 - q^2 = 5$ and $p^2 + q^2 = 13$.

14 Find positive integers p and q such that $p^2 - q^2 = 7$, $p^2 + q^2 = 25$, and use these values of p and q to construct a Pythagorean triple.

15 If (x, y, z) is a primitive Pythagorean triple and x is even, prove that
(i) $z+y$ is even,
(ii) $z-y$ is even,
(iii) $\gcd(z+y, z-y)=2$,
(iv) $\frac{1}{2}(z+y)$ and $\frac{1}{2}(z-y)$ are squares.
Deduce by putting $p^2=\frac{1}{2}(z+y)$ and $q^2=\frac{1}{2}(z-y)$ that $x=2pq$, $y=p^2-q^2$, $z=p^2+q^2$ and p and q have no common factor.

16 If $(2pq, p^2-q^2, p^2+q^2)$ is a primitive Pythagorean triple, can both p and q be even? Can both p and q be odd?

17 If (x, y, z) is a primitive Pythagorean triple, prove that $(2xy, \pm(x^2-y^2), x^2+y^2)$ is also a primitive Pythagorean triple, and if x is even, then $(2xz, z^2-x^2, z^2+x^2)$ is another primitive Pythagorean triple.
Deduce that non-zero integers a, b, c can be found such that $a^2+b^2=c^4$, and also such that $a^2+b^4=c^2$.

18 Is there an infinity of distinct primitive Pythagorean triples?

The equation $x^4+y^4=z^4$

19

1						
16	17					
81	82	97				
256	257	272	337			
625	626	641	706	881		
1296	1297	1312	1377	1552	1921	
2401	2402	2417	2482	2657	3026	3697

In the table given here, the first column is a list of fourth powers and the other columns are obtained by adding $1^4, 2^4, \ldots, 6^4$ to the entries in the first column.
Is there a square number in this table outside the first column?

20 If the table of q 19 were extended downwards, and a square were to be found in it outside the first column, then there would be integers x, y, z such that $x^4+y^4=z^2$.
If z^2 were the smallest such square to appear in the table outside the first column, with (x^2, y^2, z) a primitive Pythagorean triple, and if x^2 were even, then there would be positive integers p and q, with $\gcd(p, q)=1$, such that $x^2=2pq$, $y^2=p^2-q^2$, $z=p^2+q^2$, from q 15. Use q 16 to deduce that (q, y, p) is a primitive Pythagorean triple with q even and p odd. Now, from q 15 again, there are integers a and b, with $\gcd(a, b)=1$ such that $q=2ab$,

$y = a^2 - b^2$ and $p = a^2 + b^2$. Can any pair of the three numbers a, b, $a^2 + b^2$ have a common factor? Express x^2 in terms of a and b and deduce that a, b and $a^2 + b^2$ are all squares. Deduce that p would appear as a smaller square than z^2 in the table of sums of fourth powers.

21 The contradiction of q 20 establishes that there are no non-zero integers x, y, z satisfying $x^4 + y^4 = z^2$. Deduce that there are no non-zero integers such that $x^4 + y^4 = z^4$, or indeed $x^{4m} + y^{4m} = z^{4m}$, for any integer m.

22 If there are positive integers x, y, z such that

$$x^4 + y^2 = z^4,$$

we consider the primitive Pythagorean triple (x^2, y, z^2) with smallest z. If in this triple y is even, use the equations $y = 2pq$, $x^2 = p^2 - q^2$, $z^2 = p^2 + q^2$ to obtain an equation $q^4 + (xz)^2 = p^4$, contradicting the minimality of z.

23 If (x^2, y, z^2) is a primitive Pythagorean triple and y is odd, use the equations $x^2 = 2pq$, $y = p^2 - q^2$, $z^2 = p^2 + q^2$ to prove that there are positive integers a and b with no common factor such that $\{p, q\} = \{2ab, a^2 - b^2\}$. Can any pair of the three numbers a, b, $a^2 - b^2$ have a common factor? Express x^2 in terms of a and b and deduce that all these three numbers are squares. Compare the equation

$$b^2 + \left(\frac{x^2}{4ab}\right) = a^2$$

with the equation $x^4 + y^2 = z^4$, to contradict again the minimality of z in q 22.

The equation $x^2 + y^2 + z^2 = t^2$

24 By examining the possibilities modulo 4, find out whether there can exist integers x, y, z, t such that $x^2 + y^2 + z^2 = t^2$
(i) when x, y, and z are all odd numbers,
(ii) when x and y are odd numbers and z is even.

25 If x, y, z, and t are integers such that $x^2 + y^2 + z^2 = t^2$, prove that at least one of these four numbers is divisible by 3.

26 If x, y, z, and t are positive integers such that $x^2 + y^2 + z^2 = t^2$ and both x and y are even numbers, prove that z and t are either both even, or both odd. Deduce that $t + z$ and $t - z$ are both even

numbers. Letting $l=\frac{1}{2}x$, $m=\frac{1}{2}y$ and $n=\frac{1}{2}(t-z)$, prove that $n|l^2+m^2$, and by expressing z in terms of l, m and n, prove that $l^2+m^2>n^2$.
What are the two pairs of positive integers $\{z, t\}$ such that $26^2+8^2+z^2=t^2$?

The equation $x^3+y^3=z^3$

Questions 27–68 establish that non-zero integers x, y, z such that $x^3+y^3=z^3$ do not exist. The remaining chapters of the book are independent of this section.

27 Table 5.2 shows sums of cubes, arranged to avoid duplication. Are there any cubes in this table other than those in the first column? Use the first column as a checklist. $11^3=1331$, $12^3=1728$.

Table 5.2

−1000	−999	−992	−973	−936	−875	−784	−657	−488	−271	
−729	−728	−721	−702	−665	−604	−513	−385	−217		
−512	−511	−504	−485	−448	−387	−296	−169			
−343	−342	−335	−316	−279	−218	−127				
−216	−215	−208	−189	−152	−91					
−125	−124	−117	−98	−61						
−64	−63	−56	−37							
−27	−26	−19								
−8	−7									
−1										
1	2									
8	9	16								
27	28	35	54							
64	65	72	91	128						
125	126	133	152	189	250					
216	217	224	243	280	341	432				
343	344	351	370	407	468	559	686			
512	513	520	539	576	637	728	855	1024		
729	730	737	756	793	854	945	1072	1241	1458	
1000	1001	1008	1027	1064	1125	1216	1343	1512	1729	2000

28 If there were integers x, y, and z such that $x^3+y^3=z^3$, how many of x, y and z might be even?

29 If there were integers x, y and z such that $x^3+y^3=z^3$, how many of x, y and z might be divisible by 7?
Use table 3.2 to make a table of sums of cubes in $\mathbf{Z}_7$.

30 If there were integers x, y and z such that $x^3+y^3=z^3$, how many of x, y and z might be divisible by 13?
Use table 3.2 to make a table of sums of cubes in $\mathbf{Z}_{13}$.

31 If there were integers x, y and z such that $x^3 + y^3 = z^3$, explain why $x + y \equiv z \pmod 3$, and deduce that $(x + y - z)^3$ would have a factor 27. Expand the product $(x + y - z)^3$ and show that $(x + y)(x - z)(y - z)$ would have a factor 9.
Deduce that either x, or y or z would be divisible by 3.

32 If there were non-zero integers x, y and z satisfying $x^3 + y^3 = z^3$, would there be a solution with gcd $(x, y, z) = 1$?

33 Draw a regular hexagon with vertices on the circumference of a circle with radius 1. If the centre has coordinates $(0, 0)$ and one of the vertices is $(1, 0)$, find the coordinates of the other five vertices.

34 Let the vertex with negative x and positive y-coordinates in q 33 be denoted by ω. If $\omega = (a, b)$, work out $(a + ib)^2 = \omega^2$ and $(a + ib)^3 = \omega^3$ as complex numbers.
Show that the vertices of the hexagon may be appropriately labelled ± 1, $\pm\omega$, $\pm\omega^2$ on an Argand diagram.
Do these points form a group under multiplication when represented by complex numbers? Find $1 + \omega + \omega^2$.

Fig. 5.2

An isometric grid is formed by the standard tessellation of congruent equilateral triangles. For the rest of this chapter, a *lattice point* will mean a point of intersection of grid lines in an isometric grid (see fig. 5.2), or equivalently, a vertex of an equilateral triangle in the tessellation. We explore the possibility of matching the points of such a lattice with pairs of integers.

35 On an isometric grid, label the lattice points as points of an Argand diagram, by first choosing two lattice points as close together as possible and labelling them 0 and 1. Then choose the appropriate point to label ω with the conventions of q 34. Mark the lattice points

corresponding to ± 1, ± 2, ± 3, ± 4 and $\pm\omega$, $\pm 2\omega$, $\pm 3\omega$, $\pm 4\omega$. Noting that the ordinary convention of addition, both for complex numbers illustrated in an Argand diagram, and for vectors in the plane, locates $\alpha+\beta$ twice as far from 0 as the mid-point of the line joining the points α and β, so that 0, α, β and $\alpha+\beta$ form a parallelogram, determine whether all the lattice points of an isometric grid may be labelled with elements of the form $a+b\omega$, where a and b are integers.

36 Find integers a, b, c, and d such that $\omega^2=a+b\omega$ and $-\omega^2=c+d\omega$.

37 Does the set $\mathbf{Z}[\omega]=\{a+b\omega \mid a, b\in\mathbf{Z}\}$, regarded as a subset of the complex numbers, form a group under addition? Is the set $\mathbf{Z}[\omega]$ closed under multiplication? Do the non-zero elements of $\mathbf{Z}[\omega]$ form a group under multiplication?

38 Simplify the products
$(1+\omega)(1+\omega^2)$, $(a+b\omega)(a+b\omega^2)$, $(1-\omega)(1-\omega^2)$,
$(a-b\omega)(a-b\omega^2)$.

39 Factorise 3, a^2-ab+b^2, a^2+ab+b^2, a^3+b^3, a^3-b^3 in $\mathbf{Z}[\omega]$. Assume that a and b are integers.

40 Because $a+b\omega=(a-\tfrac{1}{2}b)+\mathrm{i}\tfrac{1}{2}b\sqrt{3}$, the modulus
$|a+b\omega|=\sqrt{[(a-\tfrac{1}{2}b)^2+3b^2/4]}$.
Show that $|a+b\omega|^2$ is an integer, and is positive unless $a+b\omega=0$.

41 The *norm* of $a+b\omega$, written $N(a+b\omega)=a^2-ab+b^2$, where a, $b\in\mathbf{Z}$. Justify the equation
$N(\alpha\beta)=N(\alpha)N(\beta)$ for α, $\beta\in\mathbf{Z}[\omega]$.

42 By reference to q 35, list those elements a of $\mathbf{Z}[\omega]$ for which
(i) $N(\alpha)=0$, (ii) $N(\alpha)=1$, (iii) $N(\alpha)=2$, (iv) $N(\alpha)=3$,
(v) $N(\alpha)=4$.

43 If α, $\beta\in\mathbf{Z}[\omega]$ and $\alpha\beta=1$, prove that $N(\alpha)=N(\beta)=1$, and deduce precisely which elements of $\mathbf{Z}[\omega]$ have multiplicative inverses in this set. These six elements are called the *units* of $\mathbf{Z}[\omega]$.

44 If α, β, $\gamma\in\mathbf{Z}[\omega]$ and $\alpha\beta\gamma=1$, find the ten possible unordered triples $[\alpha, \beta, \gamma]$.

45 Let $\lambda=1+2\omega$. With the conventions of q 35, mark the points $\pm\lambda$, $\pm 2\lambda$, $\pm\omega\lambda$, $\pm\omega^2\lambda$, $\pm\lambda^2$, $\pm\omega\lambda^2$, $\pm\omega^2\lambda^2$, $\pm\lambda\pm 1$, $\pm 2\lambda\pm 1$,

⟦108⟧

$\pm\omega\lambda \pm 1$, $\pm\omega^2\lambda \pm 1$ on an isometric grid. Can you draw a family of parallel lines which only pass through lattice points for multiples of λ?

46 Establish that it is not possible to find an $\alpha \in \mathbf{Z}[\omega]$ such that $\alpha\lambda = 1$, so that 1 is not a multiple of λ. Locate the lattice points on the grid which correspond to (a multiple of λ)+1. Can you draw a family of parallel lines containing only such lattice points?
Prove that there are no elements $\alpha, \beta \in \mathbf{Z}[\omega]$ such that $\alpha\lambda = \beta\lambda + 1$.

47 Show that -1 is not a multiple of λ in $\mathbf{Z}[\omega]$.
Locate the lattice points for (a multiple of λ)-1.
Prove that there are no elements $\alpha, \beta \in \mathbf{Z}[\omega]$ such that $\alpha\lambda = \beta\lambda - 1$.

48 Show that 2 is not a multiple of λ in $\mathbf{Z}[\omega]$. Deduce that there are no elements $\alpha, \beta \in \mathbf{Z}[\omega]$ such that $\alpha\lambda + 1 = \beta\lambda - 1$.

49 If a and b are integers, prove that $a + b\omega = a + b +$(a multiple of λ).
Use the fact that $\lambda^2 = -3$ to prove that if $\alpha \in \mathbf{Z}[\omega]$ then $\alpha \equiv 0$, 1 or -1 (mod λ).

50 What are the six possible pairs $\{\alpha, \beta\}$ in $\mathbf{Z}[\omega]$ such that $\alpha\beta = \lambda$?

51 If $\gamma \in \mathbf{Z}[\omega]$ is not a unit, and the only factors of γ are units and products of γ with a unit, then γ is called a *prime* in $\mathbf{Z}[\omega]$. Are 2, 3 and λ primes in $\mathbf{Z}[\omega]$?

52 With division defined as on complex numbers, find rational numbers u and v such that

$$(6+7\omega)/(2+3\omega) = u + v\omega,$$

using q 38 if you wish. By choosing U to be the integer closest in value to u, and V to be the integer closest in value to v, and letting $u = U + u'$, $v = V + v'$, verify that $(2+3\omega)(u'+v'\omega)$ is an element of $\mathbf{Z}[\omega]$, and that $N[(2+3\omega)(u'+v'\omega)] < N(2+3\omega)$.

53 If $\alpha = a + b\omega \in \mathbf{Z}[\omega]$ and $\beta = c + d\omega \in \mathbf{Z}[\omega]$, with $\beta \neq 0$, prove that $\alpha/\beta = u + v\omega$, for some rational numbers u and v. By choosing U to be the integer closest in value to u, and V to be the integer closest in value to v, and letting $u = U + u'$, $v = V + v'$, explain why $|u'|, |v'| \leq \frac{1}{2}$, and deduce that $|u'^2 - u'v' + v'^2| \leq \frac{3}{4}$.
Prove that $(c + d\omega)(u' + v'\omega) \in \mathbf{Z}[\omega]$, and that

$$N[(c+d\omega)(u'+v'\omega)] < N(c+d\omega).$$

Deduce that $\alpha = \beta q + r$, where $q = U + V\omega$, and $N(r) < N(\beta)$.

⟦109⟧

54 If α and β are given elements of $\mathbf{Z}[\omega]$, is the set $A=\{x\alpha+y\beta \mid x, y\in\mathbf{Z}[\omega]\}$ a subgroup of $(\mathbf{Z}[\omega], +)$? Compare the argument with that of q 1.29. Let $\delta\in A$ be an element with least non-zero norm in A, use arguments similar to those of q 1.33 and q 1.34 to show that
(i) for some x_1, y_1, $\delta=x_1\alpha+y_1\beta$,
(ii) every factor of α and β is a factor of δ,
(iii) both α and β are multiples of δ, using q 53,
(iv) A consists of all multiples of δ.
We write $\delta=\gcd(\alpha, \beta)$.

55 If π, α, $\beta\in\mathbf{Z}[\omega]$, π is prime and $\pi|\alpha\beta$, but π does not divide α, what is the subgroup of $(\mathbf{Z}[\omega], +)$ given by $A=\{x\pi+y\alpha \mid x, y\in\mathbf{Z}[\omega]\}$?
Deduce that for some x and y, $x\pi+y\alpha=1$.
Modify the argument of q 1.52 to show that $\pi|\beta$.

56 If $\pi_1, \pi_2, \ldots, \pi_n$ and $\gamma_1, \gamma_2, \ldots, \gamma_m$ are primes in $\mathbf{Z}[\omega]$ such that
$$\pi_1\pi_2\ldots\pi_n=\gamma_1\gamma_2\ldots\gamma_m,$$
prove that π_1 is equal to one of the γ_i times a unit. After dividing through by this π_1, prove that π_2 is equal to another of the γ_i times a unit. By an induction on the number of factors, prove that $n=m$ and each of the π_i is equal to one of the γ_i times a unit.
(*Unique factorisation theorem* for $\mathbf{Z}[\omega]$)

57 If there were non-zero integers x, y, z satisfying $x^3+y^3=z^3$, would there be a solution with $x, y, z\in\mathbf{Z}[\omega]$?

58 If there were non-zero $x, y, z\in\mathbf{Z}[\omega]$ satisfying $x^3+y^3=z^3$, would there be non-zero $\alpha, \beta, \gamma\in\mathbf{Z}[\omega]$ satisfying
$$\alpha^3+\beta^3+\gamma^3=0?$$

59 If there were non-zero $x, y, z\in\mathbf{Z}[\omega]$ such that $x^3+y^3+z^3=0$, would there be non-zero $\alpha, \beta, \gamma\in\mathbf{Z}[\omega]$ such that $\alpha^3+\beta^3+\gamma^3=0$ with no two of α, β and γ having a common prime factor?

60 If $x=\alpha\lambda+1$, with λ as in q 45, prove that $x^3-1=\alpha(\alpha^2-1)\lambda^3-\alpha^2\lambda^4$.
Explain why either α, $\alpha-1$ or $\alpha+1$ has a factor λ and deduce that $x^3\equiv1 \pmod{\lambda^4}$.
If $x=\alpha\lambda-1$, prove that $x^3\equiv-1 \pmod{\lambda^4}$.

61 If $x^3+y^3+z^3\equiv0 \pmod 9$, prove that the supposition that neither x, nor y, nor z has a factor λ is absurd.

⟦111⟧

62 If $x^3+y^3+z^3=0$, and λ is a factor of x but not a factor of y or z, explore the possible values of $y^3+z^3 \pmod{\lambda}$, and determine whether it is possible for y and z to be congruent modulo λ.

63 If $y=\alpha\lambda+1$ and $z=\beta\lambda-1$, prove that $y+\omega z$, $\omega y+z$ and $y+z$ each have a factor λ, and that if $\delta\lambda$ is a common factor of any two of these numbers, then δ is a unit or a common factor of y and z, so at most one of these numbers has a factor λ^2.

64 If $x^3+y^3+z^3=0$ and $\lambda | x$ but λ is not a factor of y or z, prove that $x^3 \equiv 0 \pmod{\lambda^4}$ and deduce that $\lambda^2 | x$.

65 If $x^3+y^3+z^3=0$, prove that

$$-x^3=\omega^2(y+z)(\omega y+z)(y+\omega z).$$

If x, y, and z have no prime factor in common and $\lambda | x$, use q 62 and q 63 to prove

$$\frac{\omega^2(y+z)}{\lambda}, \frac{\omega y+z}{\lambda} \quad \text{and} \quad \frac{y+\omega z}{\lambda}$$

are each cubes times a unit in $\mathbf{Z}[\omega]$, and that the highest power of λ remaining in any one of them is less than the highest power of λ in x^3, and by q 64 is greater than or equal to 1.

66 (i) By considering norms, prove that there is no element $\pi \in \mathbf{Z}[\omega]$ such that

$$\pi\lambda^3 = 1+\omega, \quad 1-\omega, \quad -1+\omega \quad \text{or} -1-\omega.$$

(ii) If α, β and γ are non-zero elements of $\mathbf{Z}[\omega]$ and λ divides exactly one of α, β and γ, prove that no equation of the form

$$\alpha^3+\omega\beta^3+\omega^2\gamma^3=0$$

can hold.

67 For any y and z, show that

$$\frac{\omega^2(y+z)}{\lambda}+\frac{\omega y+z}{\lambda}+\frac{y+\omega z}{\lambda}=0.$$

Use q 65, q 44 and q 66 to prove that if $x^3+y^3+z^3=0$ with $\lambda | x$ but with λ not dividing either y or z, then an equation $\alpha^3+\beta^3+\gamma^3=0$ may be constructed in which $\lambda | \alpha$ but λ does not divide either β or γ, and the highest power of λ dividing α is less than the highest power of λ dividing x. By repeating this process establish a contradiction to q 61 and deduce that there are no non-zero integers x, y, z satisfying $x^3+y^3+z^3=0$.

68 Prove that there are no non-zero integers x, y, z such that

$$x^{3m}+y^{3m}=z^{3m}$$

for any non-zero integer m.

⟦112⟧

Notes and answers

For concurrent reading, see bibliography: Sierpinski (1962), Bolker (1970).

1 25, 100, 169, 225, 289, 400, 625 (twice), 676, 841, 900, 1156, 1369, 1521, 1681, 2500, 2704.

2 25, 100, 225, 400, 625, 900. $4^2+3^2=5^2 \Rightarrow (4n)^2+(3n)^2=(5n)^2$.

3 169, 676, 1521, 2704. $12^2+5^2=13^2 \Rightarrow (12n)^2+(5n)^2=(13n)^2$.

4 $48^2+14^2=50^2=2500$, $30^2+16^2=34^2=1156$.

5 (4, 3, 5), (12, 5, 13), (24, 7, 25), (15, 8, 17), (35, 12, 37), (21, 20, 29), (40, 9, 41).

Fig. 5.3

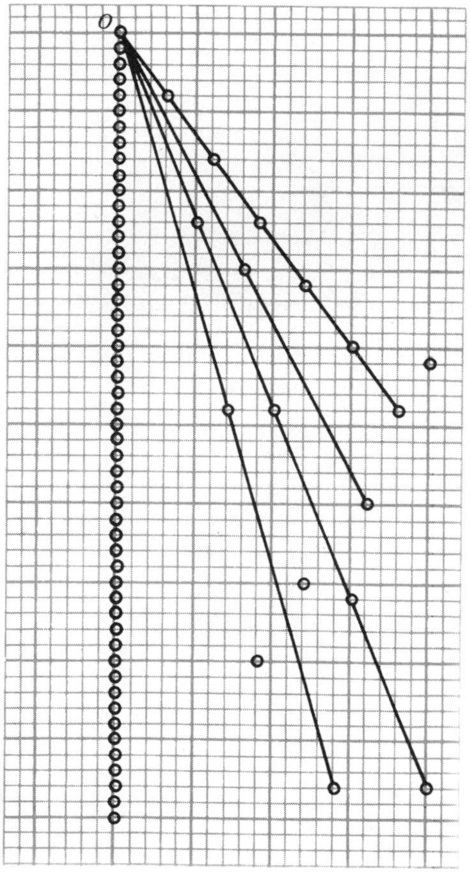

When a right-angled triangle has sides of integral length, it is said to be a Pythagorean triangle. To each Pythagorean triple there corresponds a Pythagorean triangle and vice-versa. The Pythagorean triangles implied by table 5.1 are illustrated in fig. 5.3. Each encircled lattice point is the vertex of a Pythagorean triangle with one vertex at O and two of its sides parallel to the grid lines. On a line through O there is at most one encircled lattice point corresponding to a primitive Pythagorean triple.

6 (i) $8^2+6^2=10^2$. (ii) two even imply three even. (iii) $4^2+3^2=5^2$. (iv) if x, y odd, then x^2, y^2 odd, so z^2 even, and z even. But see q 9.

7 (i) $12^2+9^2=15^2$. (ii) two divisible by 3 imply all divisible by 3. (iii) $4^2+3^2=5^2$.

(iv) Since $z^2 \not\equiv 2 \pmod 3$, either x or $y \equiv 0$ modulo 3.

x^2+y^2	0	1	2
0	0	1	1
1	1	2	2
2	1	2	2

8 (i) $20^2+15^2=25^2$. (ii) two divisible by 5 imply all divisible by 5. (iii) $4^2+3^2=5^2$.

(iv) Since $z^2 \not\equiv 2, 3 \pmod 5$, either x or y or $z \equiv 0 \pmod 5$.

x^2+y^2	0	1	2	3	4
0	0	1	4	4	1
1	1	2	0	0	2
2	4	0	3	3	0
3	4	0	3	3	0
4	1	2	0	0	2

9

x^2+y^2	0	1	2	3
0	0	1	0	1
1	1	2	1	2
2	0	1	0	1
3	1	2	1	2

Since $z^2 \not\equiv 2 \pmod 4$, z^2 even implies $z^2 \equiv 0 \pmod 4$ and this implies $x^2 \equiv y^2 \equiv 0 \pmod 4$.

Let $y=2k+1$, $z=2h+1$, then

$$\begin{aligned} x^2 &= z^2 - y^2 \\ &= 4h^2+4h-(4k^2+4k) \\ &= 4h(h+1)-4k(k+1). \end{aligned}$$

Now $h(h+1)$ and $k(k+1)$ are both even numbers, so $8|x^2$ and $4|x$.

From q 6, q 7, q 8 and q 9 we can deduce that of the three numbers forming a Pythagorean triple, at least one is divisible by 3, one by 4 and one by 5.

10 Given $x^2+y^2=z^2$,

$$\begin{aligned} &2x(y+z):(y+z)^2-x^2:(y+z)^2+x^2 \\ &\quad = 2x(y+z):2y(y+z):2z(y+z). \end{aligned}$$

12 $p=2$, $q=1$.

13 $p=3$, $q=2$.

14 $p=4$, $q=3$, (24, 7, 25).

15 From q 6, y and z are odd, so $z+y$ and $z-y$ are even.
If $\gcd(z+y,\ z-y)=2d$, then $2d|(z+y)+(z-y)$, so $d|z$ and $2d|(z+y)-(z-y)$, so $d|y$. But $\gcd(y,z)=1$, so $d=1$.
$[\frac{1}{2}(z-y)][\frac{1}{2}(z+y)]=(\frac{1}{2}x)^2$. But $\gcd(\frac{1}{2}(z+y), \frac{1}{2}(z-y))=1$, so both are squares. If $d|p, q$, then $d^2|x, y, z$, so $d=1$.

16 If p and q have the same parity, then all three members of the triple are even and the triple is not primitive.

17 By q 6 and q 9, x and y have different parity, so x^2-y^2 and x^2+y^2 are odd. If $d|x^2-y^2, x^2+y^2$, d is odd, but $d|2x^2, 2y^2$, so $d|x, y$ and $d=1$. Likewise x and z have different parity, so the same argument applies.

18 The second construction of q 17 gives an increased z, so repetition of this construction never repeats a triple.

19 No.

20 Either x^2 or y^2 must be even from q 6 and q 9, so there is no loss of generality in supposing x^2 even. From q 16, p and q have no common factor, so (q, y, p) is primitive. Since y is odd, q is even from q 6 and q 9. If $d|a, a^2+b^2$, then $d|b^2$, so $d|b$, for any prime d. But $\gcd(a,b)=1$, so $d=1$. Similarly $\gcd(b, a^2+b^2)=1$. $x^2=2(a^2+b^2)2ab$, so $(\frac{1}{2}x)^2=ab(a^2+b^2)$. Since the three factors on the right are coprime, all are squares. Thus a^2 and b^2 are fourth powers, a^2+b^2 is a square and $p=a^2+b^2$. Since $\gcd(a,b)=1$, the Pythagorean triple here is primitive, and since $p<p^2<p^2+q^2=z<z^2$, this contradicts the minimality of z.

If we had not insisted that z was the smallest such number, we should have constructed a smaller z. Fermat's 'method of descent' then provides a contradiction, for a repeated method of descent among the positive integers must terminate. The 'method of descent' is equivalent to the 'well-ordering principle' that every non-empty set of positive integers has a least member, and this in turn is equivalent to the principle of mathematical induction.

21 A Pythagorean triple of the form (x^2, y^2, z^2) would necessarily be of the form (x^2, y^2, t) and no such one exists.
A Pythagorean triple of the form (x^{2m}, y^{2m}, z^{2m}) cannot exist for positive m by the previous sentence. If m is negative and $m=-n$, then

$$x^{4m}+y^{4m}=z^{4m} \Leftrightarrow (yz)^{4n}+(zx)^{4n}=(xy)^{4n},$$

which is equally impossible.

22 Since $\gcd(p,q)=1$, (q^2, xz, p^2) is primitive and $p^2<p^2+q^2=z^2$.

23 If $d|a, a^2-b^2$, then $d|b^2$ so $d|b$ for prime d, but $\gcd(a,b)=1$. $x^2=2\cdot 2ab\cdot(a^2-b^2)$, so $(\frac{1}{2}x)^2=ab(a^2-b^2)$, and since the three factors on the right are coprime, each is a square. Now, $b^2+(x^2/4ab)=a^2$ is an equation of the form $x^4+y^2=z^4$ and since $\gcd(a,b)=1$, the Pythagorean triple here is primitive and $a<2ab<p^2+q^2=z^2$.

24 If x and y are odd numbers, $x^2 \equiv y^2 \equiv 1 \pmod 4$, so $x^2 + y^2 \equiv 2 \pmod 4$. Now if z is even, $x^2+y^2+z^2 \equiv 2 \pmod 4$, and if z is odd, $x^2+y^2+z^2 \equiv 3 \pmod 4$, but $t^2 \not\equiv 2, 3 \pmod 4$ so at least two of x, y, and z must be even.

25 If x, y, z are not divisible by 3, $x^2 \equiv y^2 \equiv z^2 \equiv 1 \pmod 3$, so $x^2+y^2+z^2 \equiv 0 \pmod 3$ and t is divisible by 3.

26 If x and y are even, $t^2 - z^2 = x^2 + y^2$ implies $t^2 \equiv z^2 \pmod 2$. Thus $t \equiv z \pmod 2$, and both $t-z$ and $t+z \equiv 0 \pmod 2$. $4l^2 + 4m^2 = (t-z)(t+z)$, so $l^2 + m^2 = n[\frac{1}{2}(t+z)]$, and $n \mid l^2+m^2$. $z = (l^2+m^2)/n - n > 0$, so $l^2+m^2 > n^2$. If $l = 13$, $m = 4$, $l^2 + m^2 = 185 = 5 \cdot 37$, so $n = 1, 5$; $z = 184, 32$; $t = 186, 42$.

27 No.

28 Three or one.

29 The only cubes in $\mathbf{Z}_7$ are 0, 1, 6. So if a sum of cubes is a cube, either just one is congruent to 0 (mod 7) or all three are congruent to 0 (mod 7).

0	1	6
1	2	0
6	0	5

30 The only cubes in $\mathbf{Z}_{13}$ are 0, 1, 5, 8, 12. So if a sum of cubes is a cube, one of the cubes is congruent to 0 (mod 13).

0	1	5	8	12
1	2	6	9	0
5	6	10	0	4
8	9	0	3	7
12	0	4	7	11

31 $x^3 \equiv x \pmod 3$, so $x + y \equiv z \pmod 3$ and $x + y - z$ has a factor 3, and $(x+y-z)^3$ has a factor 27.

$(x+y-z)^3 = (x^3+y^3-z^3) + 3(x+y)(x-z)(y-z)$.

Now 27 divides the right side and $9 \mid (x+y)(x-z)(y-z)$. It follows that $3 \mid x+y$ or $x-z$ or $y-z$.
If $3 \mid x+y$, since $3 \mid x+y-z$, it follows that $3 \mid z$. The other two possibilities lead to $3 \mid y$ or $3 \mid z$.

32 If gcd $(x, y, z) = d$, then $\left(\frac{x}{d}\right)^3 + \left(\frac{y}{d}\right)^3 = \left(\frac{z}{d}\right)^3$.

33 Vertices are $(\pm 1, 0)$, $(\pm\frac{1}{2}, \pm\frac{1}{2}\sqrt{3})$. Here we begin to construct a context in which the equation $x^3 + y^3 = z^3$ can be analysed more readily than in $\mathbf{Z}$.

34 $\omega^2 = -\frac{1}{2} - i\frac{1}{2}\sqrt{3}$, $\omega^3 = 1$. Group of order 6 generated by $-\omega$.

36 $\omega^2 = -1 - \omega$, $-\omega^2 = 1 + \omega$.

37 Yes. Yes. No multiplicative inverse for 2 in $\mathbf{Z}[\omega]$.

38 1, a^2-ab+b^2, 3, a^2+ab+b^2.

39 $3=(1-\omega)(1-\omega^2)$, $a^2-ab+b^2=(a+b\omega)(a+b\omega^2)$,
$a^2+ab+b^2=(a+b\omega)(a-b\omega^2)$, $a^3+b^3=(a+b)(a+b\omega)(a+b\omega^2)$,
$a^3-b^3=(a-b)(a-b\omega)(a-b\omega^2)$.
It is because x^3+y^3 has linear factors in $\mathbf{Z}[\omega]$ that this is an appropriate context in which to analyse the equation $x^3+y^3=z^3$.

40 $(a-\frac{1}{2}b)^2+\frac{3}{4}b^2=a^2-ab+\frac{1}{4}b^2+\frac{3}{4}b^2$

$$=a^2-ab+b^2\in\mathbf{Z}.$$

$(a-\frac{1}{2}b)^2+\frac{3}{4}b^2$ is plainly positive or zero and only zero if both parts zero, that is when $b=0$, and then $a=0$.

41 $N(\alpha)=|\alpha|^2$. For any complex numbers, α, β, $|\alpha\beta|=|\alpha|\cdot|\beta|$. $N(\alpha\beta)=|\alpha\beta|^2=|\alpha|^2\cdot|\beta|^2=N(\alpha)N(\beta)$.

42 Draw circles centre 0, radii 0, 1, $\sqrt{2}$, $\sqrt{3}$, 2. $N(\alpha)=0\Rightarrow\alpha=0$. $N(\alpha)=1\Rightarrow\alpha=\pm1$, $\pm\omega$, $\pm\omega^2$. There is no α such that $N(\alpha)=2$. $N(\alpha)=3\Rightarrow\alpha=(1+2\omega)$ times ±1, $\pm\omega$, $\pm\omega^2$. $N(\alpha)=4\Rightarrow\alpha=2$ times ±1, $\pm\omega$, $\pm\omega^2$.

43 If $\alpha\beta=1$, then $N(\alpha\beta)=N(1)$, so $N(\alpha)N(\beta)=1$. But $N(\alpha)$ and $N(\beta)$ are positive integers or zero, so $N(\alpha)=N(\beta)=1$. Thus only the six elements ±1, $\pm\omega$, $\pm\omega^2$ can have multiplicative inverses. The units in $\mathbf{Z}$ are ±1.

44 $\alpha\beta\gamma=1\Rightarrow N(\alpha)N(\beta)N(\gamma)=1$

$$\Rightarrow N(\alpha)=N(\beta)=N(\gamma)=1$$

$$\Rightarrow \alpha,\beta,\gamma=\pm1,\pm\omega,\pm\omega^2.$$

$[\alpha,\beta,\gamma]=[1,1,1],[1,-1,-1],[\omega,\omega,\omega],$
$[\omega^2,\omega^2,\omega^2],[\omega,-\omega,-\omega],[\omega^2,-\omega^2,-\omega^2],$
$[1,\omega,\omega^2],[1,-\omega,-\omega^2],[-1,\omega,-\omega^2],$
$[-1,-\omega,\omega^2].$

45 See fig. 5.4.

46 $N(\alpha\lambda)=1\Rightarrow N(\alpha)N(\lambda)=1$. But $N(\lambda)=3$, so no such α exists. If $\alpha\lambda=\beta\lambda+1$, $(\alpha-\beta)\lambda=1$, but there is no such $\alpha-\beta$.

47 $\alpha\lambda=-1\Rightarrow N(\alpha)N(\lambda)=1$ as above. If $\alpha\lambda=\beta\lambda-1$, $(\beta-\alpha)\lambda=1$, but there is no such $\beta-\alpha$.

48 $\alpha\lambda=2\Rightarrow N(\alpha)N(\lambda)=N(2)\Rightarrow N(\alpha)\cdot3=4$, so no such α exists. If $\alpha\lambda+1=\beta\lambda-1$, then $(\beta-\alpha)\lambda=2$, but there is no such $\beta-\alpha$.

49 $(a+b\omega)-(a+b)=b(\omega-1)=-\lambda b\omega^2$.
Now $a+b$ is an integer and so is congruent to 0, 1, -1 (mod 3).

50 $\{1,\lambda\}$, $\{-1,-\lambda\}$, $\{\omega,\omega^2\lambda\}$, $\{-\omega,-\omega^2\lambda\}$, $\{\omega^2,\omega\lambda\}$, $\{-\omega^2,-\omega\lambda\}$.

Fig. 5.4

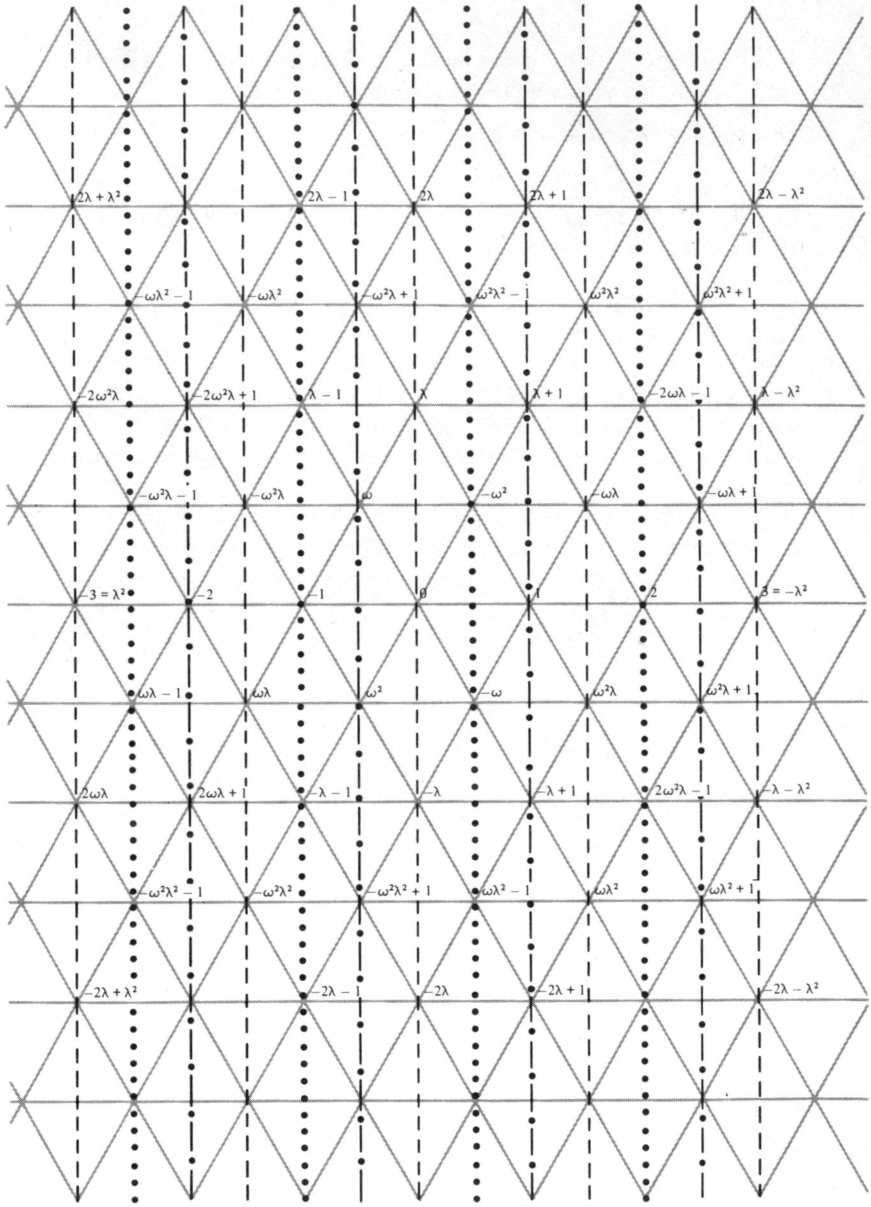
2λ + λ²
2λ − 1
2λ
2λ + 1
2λ − λ²
−ωλ² − 1
−ωλ²
−ω²λ + 1
ω²λ² − 1
ω²λ²
ω²λ² + 1
−2ω²λ
−2ω²λ + 1
λ − 1
λ
λ + 1
−2ωλ − 1
λ − λ²
−ω²λ − 1
−ω²λ
ω
−ω²
−ωλ
−ωλ + 1
−3 = λ²
−2
−1
0
1
2
3 = −λ²
ωλ − 1
ωλ
ω²
−ω
ω²λ
ω²λ + 1
2ωλ
2ωλ + 1
−λ − 1
−λ
−λ + 1
2ω²λ − 1
−λ − λ²
−ω²λ² − 1
−ω²λ²
−ω²λ² + 1
ωλ² − 1
ωλ²
ωλ² + 1
−2λ + λ²
−2λ − 1
−2λ
−2λ + 1
−2λ − λ²

51 If $\alpha\beta=2$, $N(\alpha)N(\beta)=4$. But there is no α for which $N(\alpha)=2$. So either $N(\alpha)$ or $N(\beta)=1$ and either α or β is a unit. Thus 2 is a prime. $3=(-\lambda)\lambda$ and 3 is not a prime. From q 50, λ is a prime.

52 $\dfrac{6+7\omega}{2+3\omega}=\dfrac{6+7\omega}{2+3\omega}\dfrac{2+3\omega^2}{2+3\omega^2}=\dfrac{33+14\omega+18\omega^2}{7}=\dfrac{15-4\omega}{7}.$

$U=2$, $V=-1$; $(2+3\omega)(\frac{1}{7}+\frac{3}{7}\omega)=-1$; $N(-1)<7=N(2+3\omega)$.

53 Multiply numerator and denominator by $c+d\omega^2$ and simplify. If $u=2\frac{1}{2}$, for example, there is ambiguity in the choice of U, but either $U=2$ or $U=3$ will suffice. $|u'^2-u'v'+v'^2|\leqslant u'^2+|u'v'|+v'^2\leqslant\frac{3}{4}$.
$\alpha=\beta(U+V\omega)+(c+d\omega)(u'+v'\omega)$ and $\alpha-\beta(U+V\omega)\in\mathbf{Z}[\omega]$.
$N[(c+d\omega)(u'+v'\omega)]\leqslant\frac{3}{4}N(c+d\omega)<N(c+d\omega)$, by reinterpreting the argument of q 41.
This is the division algorithm in $\mathbf{Z}[\omega]$ which we use to establish unique factorisation.

54 (iii) By q 53, $\alpha=\delta q+r$, where $N(r)<N(\delta)$. But $r=\alpha-(x_1\alpha+y_1\beta)q=(1-x_1q)\alpha+(-y_1q)\beta\in A$. Since δ has minimal positive norm in A, $N(r)=0$. So $r=0$ and α is a multiple of δ.
On the isometric grid, the set A appears as an isometric sublattice constructed from δ rather than from 1.
If $N(\gamma)=1$, then $N(\gamma\delta)=N(\delta)$, so that there are at least six elements of $\mathbf{Z}[\omega]$ satisfying the conditions for gcd (α,β). If $N(\delta)=N(\eta)$ and $\eta=\gamma\delta$, then $N(\gamma)=1$ and so there are exactly six elements satisfying the conditions for gcd (α,β).

55 Since the only divisors of π are units, π and units times π, and since π does not divide α, gcd $(\pi,\alpha)=\delta=$ a unit.
The multiples of a unit include 1, so $A=\mathbf{Z}[\omega]$, and there exist x and y such that $x\pi+y\alpha=1$, so $x\pi\beta+y\alpha\beta=\beta$. But $\pi|\alpha\beta$, so $\pi|\beta$.

56 Since $\pi_1|\pi_1\pi_2\ldots\pi_n$, $\pi_1|\gamma_1$ or $\pi_1|\gamma_2\ldots\gamma_m$ by q 55. If $\pi_1|\gamma_1$, $\pi_1=\gamma_1$ times a unit. If not, the argument may be repeated until $\pi_1|\gamma_i$, if necessary $m-1$ times. Similarly for π_2. If products of $n-1$ prime factors are unique up to units, the first step here provides the basis for an induction. The awkwardness of the units here is unavoidable, just as it would be if we stated a theorem analogous to q 1.58 for unique factorisation in $\mathbf{Z}$.

57 Yes, because $\mathbf{Z}\subset\mathbf{Z}[\omega]$.

58 Yes, let $\alpha=x$, $\beta=y$, $\gamma=-z$.

59 If gcd $(x,y,z)=d$, let $\alpha=x/d$, $\beta=y/d$, $\gamma=z/d$.

60 $x^3-1=(\alpha\lambda+1)^3-1=\alpha^3\lambda^3+3\alpha^2\lambda^2+3\alpha\lambda$. Now $3=-\lambda^2$, so $x^3-1=\alpha^3\lambda^3-\alpha\lambda^3-\alpha^2\lambda^4$. From q 49, $\alpha\equiv0, 1, -1 \pmod{\lambda}$, so $\alpha, \alpha+1, \alpha-1\equiv0 \pmod{\lambda}$. Thus $\alpha(\alpha-1)(\alpha+1)\equiv0 \pmod{\lambda}$, and so $(\alpha^3-\alpha)\lambda^3\equiv0 \pmod{\lambda^4}$. It now follows that $x^3-1\equiv0 \pmod{\lambda^4}$ and a similar argument holds if $x=\alpha\lambda-1$.

61 If x, y, $z \equiv \pm 1 \pmod{\lambda}$, then x^3, y^3, $z^3 \equiv \pm 1 \pmod{\lambda^4}$ by 60, so $x^3+y^3+z^3 \equiv \pm 3, \pm 1 \pmod{\lambda^4}$. But $\lambda^4 = 9$, so $x^3+y^3+z^3 \equiv 0 \pmod 9$ is absurd unless at least one of x, y and z has a factor λ.

62 If $y \equiv 1 \pmod{\lambda}$, then $y^3 \equiv 1 \pmod{\lambda}$. If $y \equiv -1 \pmod{\lambda}$, then $y^3 \equiv -1 \pmod{\lambda}$. Thus $y^3+z^3 \equiv \pm 1 \pm 1 \pmod{\lambda}$. But $y^3+z^3 = (-x)^3$ and $\lambda | x$, so $y^3+z^3 \equiv 0 \pmod{\lambda}$. Thus $\{y^3, z^3\} \equiv \{1, -1\}$ and so $\{y, z\} \equiv \{1, -1\}$. That is, y and z cannot be congruent modulo λ.

63 $y+\omega z = \alpha\lambda + 1 + \omega(\beta\lambda - 1) = \alpha\lambda + \beta\lambda + \omega^2\lambda$, $\omega y + z = \omega(\alpha\lambda + 1) + \beta\lambda - 1 = \alpha\lambda + \beta\lambda - \omega^2\lambda$, $y + z = \alpha\lambda + \beta\lambda$. Thus $\lambda | y+\omega z$, $\omega y + z$, $y+z$.
If $\delta\lambda | y+\omega z$ and $y+z$, then $\delta\lambda | (1-\omega)z$, and $\delta | \omega^2 z$. Thus δ is a unit or $\delta | z$. Since $\delta | y+z$, we then have $\delta | y$, z.
A similar argument holds if $\delta\lambda | \omega y + z$, $y+z$.
If $\delta\lambda | y+\omega z$, $\omega y+z$, then $\delta\lambda | y+\omega z - \omega^2(\omega y + z)$, so $\delta\lambda | \lambda z$, and $\delta | z$. Since $\delta | \omega y + z$, $\delta | \omega y$, so δ is a unit or a common factor of y and z.
λ is not a factor of y or z, so δ cannot have a factor λ, thus at most one of these three numbers has a factor λ^2.

64 From q 62, y, $z \equiv 1, -1 \pmod{\lambda}$, so from q 61, y^3, $z^3 \equiv 1, -1 \pmod{\lambda^4}$, and $-x^3 \equiv 0 \pmod{\lambda^4}$. Since $x \equiv 0 \pmod{\lambda}$, $x = \gamma\lambda$ for some γ, and so $x^3 = \gamma^3\lambda^3$. But $\lambda^4 | \gamma^3\lambda^3$, so $\lambda | \gamma^3$ and since λ is a prime $\lambda | \gamma$, so $\lambda^2 | x$.

65 For any y, z, $y^3+z^3 = \omega^2(y+z)(\omega y+z)(y+\omega z)$. From q 59 and q 62 there is no loss of generality in the hypotheses of q 63 and the supposition that y and z have no common factor other than a unit, and so no two of

$$\frac{\omega^2(y+z)}{\lambda}, \frac{\omega y + z}{\lambda}, \frac{y+\omega z}{\lambda}$$

have a common factor other than a unit, so all three are cubes, or cubes times a unit, being factors of $(-x/\lambda)^3$. From q 63 at most one of the numbers

$$\frac{\omega^2(y+z)}{\lambda}, \frac{\omega y + z}{\lambda}, \frac{y+\omega z}{\lambda}$$

has a factor λ and since the product of these numbers is $(-x/\lambda)^3$, the highest power of λ to appear as a factor in any of them is λ^{k-3} where λ^k is the highest power of λ to divide x^3.

66 (i) $N(1+\omega) = N(-1-\omega) = 1$, $N(1-\omega) = N(-1+\omega) = 3$. $N(\lambda) = 3$, so $N(\pi\lambda^3) = N(\pi) \cdot 27 \neq 1, 3$.
(ii) From q 60,

$\alpha \equiv 0 \pmod{\lambda} \Rightarrow \alpha^3 \equiv 0 \pmod{\lambda^3}$,

$\alpha \equiv 1 \pmod{\lambda} \Rightarrow \alpha^3 \equiv 1 \pmod{\lambda^3}$,

$\alpha \equiv -1 \pmod{\lambda} \Rightarrow \alpha^3 \equiv -1 \pmod{\lambda^3}$.

If $\lambda|\gamma$, but λ is not a factor of α or β, then $\alpha^3+\omega\beta^3+\omega^2\gamma^3=0$ implies $\pm1\pm\omega\equiv0 \pmod{\lambda^3}$ which is impossible by (i).
If $\lambda|\alpha$, multiply the equation through by ω^2 and then apply the same argument.
If $\lambda|\beta$, multiply the equation through by ω and then apply the same argument.

67 Since the sum of the three numbers is zero and their product is a cube, the equation given is either of the type $\alpha^3+\beta^3+\gamma^3=0$ or of the type $\alpha^3+\omega\beta^3+\omega^2\gamma^3=0$ by q 44. From q 65 exactly one of α^3, β^3 and γ^3 has a factor λ, so by q 66, $\alpha^3+\omega\beta^3+\omega^2\gamma^3=0$ is not possible. Thus from an equation of the type $x^3+y^3+z^3=0$, in which x, y and z have no common factor, just one has a factor λ, another such equation may be constructed with a lower power of λ in its term with a factor λ. By repeating this construction a sufficient number of times an equation of this type may be constructed without a term with a factor λ. This contradicts q 61. This is another example of the use of Fermat's method of descent.

68 The given equation is equivalent to $(x^m)^3+(y^m)^3=(z^m)^3$. If m is negative $=-n$, it is equivalent to $(y^nz^n)^3+(x^nz^n)^3=(x^ny^n)^3$.
The claim that there are no non-zero integers x, y, z such that $x^n+y^n=z^n$ for $n>2$ is known as Fermat's last theorem. No proof is known. Considerable effort has been expended on establishing the result for various particular prime numbers n.

Historical note

A systematic method of constructing Pythagorean triples was known to the Babylonians (c. 1500 B.C.). Our formula (5.11) was known to Euclid (c. 300 B.C.), and Diophantus (c. A.D. 200) could prove that it was complete. The non-existence of integral solutions of $x^4+y^4=z^4$ was claimed by Fermat, and his method of descent, as we have seen, provides a proof. Fermat also claimed that he had a proof that $x^3+y^3=z^3$ has no integral solutions. In 1770, L. Euler published what was long regarded as a complete proof of this fact. C. F. Gauss found a proof using complex cube roots of unity.

In the margin of his copy of Diophantus, Fermat wrote 'it is impossible to separate a cube into two cubes, or a biquadrate [i.e. fourth power] into two biquadrates, or in general any power higher than the second into two powers of like degree; I have discovered a truly remarkable proof which this margin is too small to contain'. The proposition that $x^n+y^n=z^n$ has no solution in integers, when $n>2$, is the only result claimed by Fermat for which no proof is known, and is therefore called 'Fermat's last theorem'. A complete proof would follow if Fermat's last theorem were established for all odd primes n and $n=4$. The attempt to extend the set of prime numbers, n, for which Fermat's theorem was known to hold led Kummer and others in the

1840s to identify unique factorisation domains, and the surprising absence of unique factorisation in some complex arithmetic. To date, Fermat's last theorem has only been established for $n = 4$ and what may be only a finite set of primes n.

An interesting collection of extracts from letters of Fermat have been reprinted in Bell (1962). A detailed historical survey of work done on the last theorem is given in Edwards (1977).

6

Sums of squares

Sums of two squares

1 Determine the positions in table 1.1 of the integers less than 100 which appear in table 5.1.

2 Which columns of table 1.1 contain numbers from table 5.1 and which do not?

3 Which columns of table 1.1 contain odd prime numbers from table 5.1 and which do not?

4 Make a table of sums of squares modulo 4 and prove that no integer of the form $4k+3$ can be a sum of two squares.

5 Make a list of those numbers less than 200 in tables 5.1 which have a factor 3. Determine, in each case, the highest power of 3 which divides the number.

6 Make a table of sums of squares modulo 3. What can you deduce about x and y if $x^2+y^2 \equiv 0 \pmod 3$? And what does this imply about the number x^2+y^2?

7 Make a list of those numbers less than 200 in table 5.1 which have a factor of 7. Determine, in each case, the highest power of 7 which divides the number.

8 Make a table of sums of squares modulo 7. What can you deduce about x and y if $x^2+y^2 \equiv 0 \pmod 7$? What does this imply about the number x^2+y^2?

9 Suppose a prime number p, different from 2, divides a number of the form x^2+y^2, so that $x^2+y^2 \equiv 0 \pmod p$; then if $y \not\equiv 0 \pmod p$,

there is an integer a such that $ay \equiv 1 \pmod p$, so that $(ax)^2+1 \equiv 0 \pmod p$. Thus ax is an element of order 4 in $\mathbf{M}_p$. What does Lagrange's theorem on subgroups now tell you about p? (Compare with q 4.18 and q 4.19.)

10 Try formulating a contrapositive to q 9 with a view to generalising q 6 and q 8.

11 Explore the validity of the argument in q 9 when $p=2$.

12 If x^2+y^2 is divisible by 27, must it be divisible by 81? Generalise your argument.

13 Examine the prime factorisation of numbers in table 5.1 with a view to making a conjecture about which positive integers may be expressed as a sum of two squares.

14 Find three numbers in table 5.1, each of which appears at least twice outside the first column. Express each of these numbers as a product of two numbers in the table. By using a factorisation into complex numbers of the form

$$13=2^2+3^2=(2+3\mathrm{i})(2-3\mathrm{i}),$$

express each of your three original numbers as a product of four complex numbers. By changing the order in the product of the four complex numbers, provide a rationale for the three examples of the equation $a^2+b^2=c^2+d^2$ that you have found.

15 When a and b are integers, a complex number of the form $a+ib$ is called a *Gaussian integer*.
Factorise a^2+b^2 into a product of Gaussian integers. If a, b, c, and d are integers, express $(a^2+b^2)(c^2+d^2)$ as a product of four Gaussian integers. Arrange these four factors so that the first pair consists of the complex conjugates of the second pair. Multiply the first pair together to form a Gaussian integer, and do the same with the second pair. Are the two Gaussian integers that are formed in this way complex conjugates?
Express $(a^2+b^2)(c^2+d^2)$ as the sum of two squares.

16 How does the identity found at the end of q 15 enable you to establish part of your conjecture in q 13?

17 Identify what part of your conjecture in q 13 remains unproved.

18 Find the prime factors of all the numbers in the second column of table 5.1.

⟦122⟧

19 Using q 4.22, determine whether each prime number congruent to 1 (mod 4) would occur as a factor of a number in the second column of table 5.1 if only it was extended far enough.

20 Make a list of the prime numbers of the form n^2+1 which appear in table 5.1.

21 Make a list of those numbers in table 5.1 which are twice an odd prime and have the form n^2+1.

22 The equation $35^2+1=2\cdot 613$ leads to an expression for 613 as a sum of two squares as follows.

$$(35^2+1^2)(1^2+1^2)=2^2\cdot 613$$
$$\Rightarrow (35+1)^2+(35-1)^2=2^2\cdot 613 \text{ by q 15}$$
$$\Rightarrow 36^2+34^2=2^2\cdot 613$$
$$\Rightarrow 18^2+17^2=613.$$

Apply this method to each of your equations in q 21 to express 13, 41, 61, 113, 181 and 313 as the sum of two squares.
Given that $79^2+1=2\cdot 3121$ and $85^2+1=2\cdot 3613$, express each of the prime numbers 3121 and 3613 as the sum of two squares.

23 Given that

$$42^2+1=5\cdot 353,$$
$$48^2+1=5\cdot 461,$$
$$59^2+1=5\cdot 541,$$
$$58^2+1=5\cdot 673,$$

adapt the method of q 22 to express 353, 461, 541 and 673 as the sum of two squares.

24 Adapt the method of q 23 to derive an expression for 13 as the sum of two squares from $7^2+4^2=5\cdot 13$, and an expression for 17 as the sum of two squares from $9^2+2^2=5\cdot 17$.

25 Given that $33^2+1=10\cdot 109$ and $67^2+1=10\cdot 449$, use the methods of q 22 and q 24 to express 109 and 449 as the sum of two squares. Although the work can be done in one step using $3^2+1=10$, it is more illuminating to do it in two steps, first removing a factor 2, then a factor 5.

26 Apply the two-step method of q 25 to the equation $11^2+13^2=10\cdot 29$ to derive an expression for 29 as the sum of two squares.

27 Given that $34^2+1=13\cdot 89$, find integers x and y such that $(34^2+1^2)(x^2+y^2)=13^2\cdot 89$ with $34x+y$ and $34y-x$ both divisible by 13. Hence derive an expression for 89 as the sum of two squares.

⟦124⟧

28 If p is a prime number such that $n^2+1=pk$, and $p=x^2+y^2$, prove that $nx\equiv\pm y \pmod p$, and use the equations

$$(n^2+1)(x^2+y^2)=(nx+y)^2+(ny-x)^2$$
$$=(nx-y)^2+(ny+x)^2$$

to derive an expression for k as the sum of two squares.

29 Given that $113^2+22^2=29\cdot 457$, find integers x and y such that $(113^2+22^2)(x^2+y^2)=29^2\cdot 457$ with $113x+22y$ and $113y-22x$ both divisible by 29. Hence derive an expression for 457 as the sum of two squares.

30 If p is a prime number such that $m^2+n^2=pk$, and $p=x^2+y^2$, prove that $mx\equiv\pm ny \pmod p$, and use the equations

$$(m^2+n^2)(x^2+y^2)=(mx+ny)^2+(my-nx)^2$$
$$=(mx-ny)^2+(my+nx)^2$$

to derive an expression for k as the sum of two squares.

31 How many prime factors greater than m can the number m^2+1 have?

32 If every prime number which divides a number of the form n^2+1, for $n<m$, may be expressed as the sum of two squares, prove that every prime factor of m^2+1 which is less than m can be expressed as the sum of two squares.
By repeated use of q 30, and by an appeal to q 31, prove that every prime factor of m^2+1 can be expressed as the sum of two squares.

33 Use q 32 to establish by induction that every prime factor of a number of the form n^2+1 can be expressed as the sum of two squares. Deduce that every prime number congruent to 1 (mod 4) can be expressed as the sum of two squares.

34 Establish a complete classification of those integers which can be expressed as the sum of two square integers.

35 Does any prime number appear twice in table 5.1?

36 If, for a prime number p, $p=a^2+b^2=c^2+d^2$ and $p|n^2+1$,
(i) prove that $na\equiv\pm b$ and $nc\equiv\pm d \pmod p$,
(ii) use the equations

$$p^2=(a^2+b^2)(c^2+d^2)=(ac+bd)^2+(ad-bc)^2$$
$$=(ac-bd)^2+(ad+bc)^2$$

to show that in one of the expressions on the right, both terms are divisible by p and so one must be 0.
Deduce that $\{a^2, b^2\} = \{c^2, d^2\}$.

Sums of four squares

37 Express each integer from 1 to 20 as the sum of at most four squares.

38 If the quaternion $\alpha = a + b\mathrm{i} + c\mathrm{j} + d\mathrm{k}$, where a, b, c and d are integers, and

$$\mathrm{i}^2 = \mathrm{j}^2 = \mathrm{k}^2 = -1, \qquad \mathrm{ij} = -\mathrm{ji} = \mathrm{k},$$
$$\mathrm{jk} = -\mathrm{kj} = \mathrm{i}, \qquad \mathrm{ki} = -\mathrm{ik} = \mathrm{j},$$

we define the conjugate quaternion $\bar{\alpha} = a - b\mathrm{i} - c\mathrm{j} - d\mathrm{k}$. Evaluate the product $\alpha\bar{\alpha}$ assuming the usual distributive laws, and that integers with i, j and k under multiplication.
For any two quaternions α, β, prove that $\overline{\beta\alpha} = \bar{\alpha}\bar{\beta}$.

39 For integers a, b, c, d, x, y, z, t, express

$$(a^2 + b^2 + c^2 + d^2)(x^2 + y^2 + z^2 + t^2)$$

as a product of four quaternions, and hence as a product of two conjugate quaternions.
Deduce that if two integers can be expressed as the sum of four squares, then their product can also be expressed as the sum of four squares.

40 Having already shown, in q 34, that any positive integer can be expressed as a sum of two squares, unless in its prime factorisation a prime number $p \equiv 3 \pmod 4$ occurs to an odd power, what more needs to be established if it is to be shown that every positive integer can be expressed as the sum of four squares?
(*Lagrange's four square theorem*)

41 If p is an odd prime, how many squares are there in $\mathbf{Z}_p$?
How many distinct elements are there in the set $\{x^2 + 1 \mid x \in \mathbf{Z}_p\}$ and how many in the set $\{-(x^2) \mid x \in \mathbf{Z}_p\}$?
Deduce that these two sets overlap and that there are integers x and y such that $x^2 + y^2 + 1 \equiv 0 \pmod p$.

42 If p is an odd prime, must there be integers x, y with $0 \leqslant x, y < \frac{1}{2}p$ such that $x^2 + y^2 + 1 \equiv 0 \pmod p$?
Deduce that there exist integers a, b, c, d such that

$$a^2 + b^2 + c^2 + d^2 = mp,$$

where $m < p$.

⟦127⟧

43 $2^2+4^2+6^2+8^2=120,$

$1^2+2^2+3^2+4^2=30,$

$1^2+3^2+5^2+7^2=84.$

If a, b, c and d are integers and $a^2+b^2+c^2+d^2$ is even, how many of the numbers a, b, c, and d can be odd?

44 If a and b are both even or both odd, what can be said about the numbers $a+b$ and $a-b$?

45 If

$$\left(\frac{a+b}{2}\right)^2+\left(\frac{a-b}{2}\right)^2+\left(\frac{c+d}{2}\right)^2+\left(\frac{c-d}{2}\right)^2$$
$$=k(a^2+b^2+c^2+d^2)$$

find k.

46 $1^2+3^2+5^2+7^2=84.$

Use q 45 to express the number 42 as the sum of four squares, and then use q 45 again to express the number 21 as the sum of four squares.

47 If a, b, c and d are integers and

$a^2+b^2+c^2+d^2=mp,$

where m is an even number, prove that there is a multiple of p, numerically less than mp, which can be expressed as the sum of four squares.

48 $1^2+2^2+3^2+5^2=39=3\cdot 13.$

If we choose $x\equiv 1 \pmod 3$, $y\equiv 2 \pmod 3$, $z\equiv 3 \pmod 3$ and $t\equiv 5 \pmod 3$, prove that each of the five numbers

$x^2+y^2+z^2+t^2,$

$x+2y+3z+5t,$

$y-2x-3t+5z,$

$z+2t-3x-5y,$

and

$t-2z+3y-5x$

are divisible by 3. Choose the numerically least values for x, y, z and t, and express the product

$(1^2+2^2+3^2+5^2)(x^2+y^2+z^2+t^2)$

as the sum of four squares. Hence derive an expression for 13 as the sum of four squares.

⟦128⟧

49 $1^2+1^2+2^2+17^2=295=5\cdot 59$.
By choosing numerically least x, y, z and t such that $x\equiv 1 \pmod 5$, $y\equiv 1 \pmod 5$, $z\equiv 2 \pmod 5$, $t\equiv 17 \pmod 5$, and multiplying the given equation through by $x^2+y^2+z^2+t^2$, derive an expression for 59 as the sum of four squares.

50 If a, b, c, d are integers and

$$a^2+b^2+c^2+d^2=mp,$$

where m is an odd number, and numerically least x, y, z, t are chosen so that $x\equiv a \pmod m$, $y\equiv b \pmod m$, $z\equiv c \pmod m$ and $t\equiv d \pmod m$, prove that $x^2+y^2+z^2+t^2<m^2$, and that the five numbers

$$x^2+y^2+z^2+t^2,$$

$$ax+by+cz+dt,$$

$$ay-bx-ct+dz,$$

$$az+bt-cx-dy,$$

and

$$at-bz+cy-dx$$

are all divisible by m.
Deduce that for some integer m', with $0<m'<m$, $m'p$ can be expressed as the sum of four squares.

51 If p is an odd prime, deduce from q 41, q 47 and q 50, that there is no least integer $m>1$, such that mp can be expressed as the sum of four squares, and that p can therefore be expressed as the sum of four squares.

52 From q 39, q 51 and the equation $0^2+0^2+1^2+1^2=2$, prove that every positive integer can be expressed as the sum of four squares.

Sums of three squares

53 Attempt to write each of the integers from 1 to 20 as a sum of three squares.

54 Make a table for sums of squares modulo 8.
Is it possible to find integers x, y and z such that $x^2+y^2+z^2\equiv 7 \pmod 8$?
If $x^2+y^2+z^2\equiv 0 \pmod 4$, what can be said about the numbers x^2, y^2 and z^2?
If $4m$ is the sum of three squares, prove that m is also the sum of three squares.
Prove that for no integers h, k, positive or zero, can the number $4^h(8k+7)$ be expressed as the sum of three squares.

[[129]]

Notes and answers

For concurrent reading see bibliography: Bolker (1970), Davenport (1978), Ore (1948), Shanks (1978).

1 0, 4, 8, 16, 20, 32, 36, 40, 52, 64, 68, 72, $80 \equiv 0 \pmod 4$.
1, 5, 9, 13, 17, 25, 29, 37, 41, 45, 49, 61, 65, 73, 81, 85, 89, $97 \equiv 1 \pmod 4$.
10, 18, 26, 34, 50, 58, 74, 82, 90, $98 \equiv 2 \pmod 4$.

2 All but those congruent to 3 (mod 4).

3 Only those congruent to 1 (mod 4).

4 $x^2 = x^2 + 0^2$ counts as a sum of squares, here, and throughout this chapter.

	0	1	2	3
0	0	1	0	1
1	1	2	1	2
2	0	1	0	1
3	1	2	1	2

5 9, $18 = 9 \cdot 2$, $36 = 9 \cdot 4$, $45 = 9 \cdot 5$, $72 = 9 \cdot 8$, 81, $90 = 9 \cdot 10$, $117 = 9 \cdot 13$, $144 = 9 \cdot 16$, $153 = 9 \cdot 17$, $162 = 81 \cdot 2$, $180 = 9 \cdot 20$.

6

	0	1	2
0	0	1	1
1	1	2	2
2	1	2	2

$$x^2 + y^2 \equiv 0 \pmod 3$$
$$\Rightarrow x \equiv y \equiv 0 \pmod 3$$
$$\Rightarrow x^2 + y^2 \equiv 0 \pmod 9.$$

7 49, $98 = 49 \cdot 2$, $196 = 49 \cdot 4$.

8 $0^2 \equiv 0$, $1^2 \equiv 1$, $2^2 \equiv 4$, $3^2 \equiv 2$, $4^2 \equiv 2$, $5^2 \equiv 4$, $6^2 \equiv 1 \pmod 7$

0	1	2	4
1	2	3	5
2	3	4	6
4	5	6	1

$$x^2 + y^2 \equiv 0 \pmod 7$$
$$\Rightarrow x \equiv y \equiv 0 \pmod 7$$
$$\Rightarrow x^2 + y^2 \equiv 0 \pmod{49}.$$

9 If $\mathbf{M}_p$ has an element of order 4, $4 | p-1$, so $p \equiv 1 \pmod 4$.

10 If $p \equiv 3 \pmod 4$, then

$$x^2 + y^2 \equiv 0 \pmod p$$
$$\Rightarrow x \equiv y \equiv 0 \pmod p$$
$$\Rightarrow x^2 + y^2 \equiv 0 \pmod{p^2}.$$

11 Since $1 \equiv -1 \pmod 2$, $(ax)^2 + 1 \equiv 0 \pmod 2$ does not imply that ax has order 4 in $\mathbf{M}_2$. $x^2 + y^2 \equiv 0 \pmod 2$ implies that x and y are both odd or

both even. In this respect, 2 is like the primes congruent to 1 (mod 4), not like those congruent to 3 (mod 4).

12 $27|x^2+y^2 \Rightarrow x^2+y^2 \equiv 0 \pmod 3$ so from q 6, $x=3x'$, $y=3y'$. Now $x^2+y^2=9(x'^2+y'^2)$, and $3|x'^2+y'^2$. From q 6, $x'=3x''$, $y'=3y''$, so $x^2+y^2=81(x''^2+y''^2)$. By using q 10 a similar argument establishes that if a prime $p|x^2+y^2$ and $p\equiv 3 \pmod 4$, then the highest power of p which divides x^2+y^2 is even.

13 In the prime factorisation of a number of the form x^2+y^2, prime numbers congruent to 3 (mod 4) occur to even powers, other prime numbers to any power.

14 $50=5\cdot 10=(2^2+1^2)(3^2+1^2)=(2+\mathrm{i})(2-\mathrm{i})(3+\mathrm{i})(3-\mathrm{i})$
$[(2+\mathrm{i})(3+\mathrm{i})][(2-\mathrm{i})(3-\mathrm{i})]=(5+5\mathrm{i})(5-5\mathrm{i})=5^2+5^2$
$[(2+\mathrm{i})(3-\mathrm{i})][(2-\mathrm{i})(3+\mathrm{i})]=(7+\mathrm{i})(7-\mathrm{i})=7^2+1^2$

$65=5\cdot 13=(2^2+1^2)(2^2+3^2)=(2+\mathrm{i})(2-\mathrm{i})(2+3\mathrm{i})(2-3\mathrm{i})$
$[(2+\mathrm{i})(2+3\mathrm{i})][(2-\mathrm{i})(2-3\mathrm{i})]=(1+8\mathrm{i})(1-8\mathrm{i})=1^2+8^2$
$[(2-\mathrm{i})(2+3\mathrm{i})][(2+\mathrm{i})(2-3\mathrm{i})]=(7+4\mathrm{i})(7-4\mathrm{i})=7^2+4^2$

$85=5\cdot 17=(2^2+1^2)(4^2+1^2)=(2+\mathrm{i})(2-\mathrm{i})(4+\mathrm{i})(4-\mathrm{i})$
$[(2+\mathrm{i})(4+\mathrm{i})][(2-\mathrm{i})(4-\mathrm{i})]=(7+6\mathrm{i})(7-6\mathrm{i})=7^2+6^2$
$[(2+\mathrm{i})(4-\mathrm{i})][(2-\mathrm{i})(4+\mathrm{i})]=(9+2\mathrm{i})(9-2\mathrm{i})=9^2+2^2$.

15 The Gaussian integers are denoted by $\mathbf{Z}[\mathrm{i}]$ and admit a unique factorisation theorem like q 5.56 for $\mathbf{Z}[\omega]$; this is proved by defining the norm $N(a+ib)=a^2+b^2$, but we do not use this here.

$$a^2+b^2=(a+b\mathrm{i})(a-b\mathrm{i})$$
$$(a^2+b^2)(c^2+d^2)=(a+b\mathrm{i})(a-b\mathrm{i})(c+d\mathrm{i})(c-d\mathrm{i})$$
$$\begin{aligned}[(a+b\mathrm{i})(c+d\mathrm{i})][(a-b\mathrm{i})(c-d\mathrm{i})] &= [(ac-bd)+(ad+bc)\mathrm{i}][(ac-bd)-(ad+bc)\mathrm{i}] \\ &= (ac-bd)^2+(ad+bc)^2,\end{aligned}$$

and

$$\begin{aligned}[(a+b\mathrm{i})(c-d\mathrm{i})][(a-b\mathrm{i})(c+d\mathrm{i})] &= [(ac+bd)-(ad-bc)\mathrm{i}][(ac+bd)+(ad-bc)\mathrm{i}] \\ &= (ac+bd)^2+(ad-bc)^2.\end{aligned}$$

16 The identities of q 15 show that the set of integers expressible as the sum of two squares is closed under multiplication. Thus every positive integer in whose prime factorisation every prime occurs to an even power is certainly expressible as the sum of two squares since each integer mentioned is a square, and certainly, $x^2=x^2+0^2$.

17 We still need to show that every prime congruent to 1 (mod 4) may be expressed as the sum of two squares.

18 From q 4.24 it is only necessary to search for prime factors congruent to 1 (mod 4), and from q 1.47 only a knowledge of these primes up to 41 is required. Or use Hubbard (1975).

See n 20 and n 21. In addition we have the following.
Two primes: $8^2+1=5\cdot 13$, $12^2+1=5\cdot 29$, $22^2+1=5\cdot 97$, $28^2+1=5\cdot 157$, $30^2+1=17\cdot 53$, $34^2+1=13\cdot 89$, $42^2+1=5\cdot 353$, $44^2+1=13\cdot 149$, $46^2+1=29\cdot 73$, $48^2+1=5\cdot 461$, $50^2+1=41\cdot 61$.
Three primes: $13^2+1=2\cdot 5\cdot 17$, $17^2+1=2\cdot 5\cdot 29$, $21^2+1=2\cdot 13\cdot 17$, $23^2+1=2\cdot 5\cdot 23$, $27^2+1=2\cdot 5\cdot 73$, $31^2+1=2\cdot 13\cdot 37$, $33^2+1=2\cdot 5\cdot 109$, $37^2+1=2\cdot 5\cdot 137$.
Three primes, one repeated: $7^2+1=2\cdot 5^2$, $18^2+1=5^2\cdot 13$, $32^2+1=5^2\cdot 41$, $38^2+1=5\cdot 17^2$, $41^2+1=2\cdot 29^2$.
Four primes: $43^2+1=2\cdot 5^2\cdot 37$, $47^2+1=2\cdot 5\cdot 13\cdot 17$.

19 For any prime p congruent to 1 (mod 4), -1 is a quadratic residue modulo p by q 4.18 or q 4.29, so there is an integer n such that $n^2\equiv -1 \pmod p$ and $p|n^2+1$.

20 Primes.
$1^2+1=2$, $2^2+1=5$, $4^2+1=17$, $6^2+1=37$, $10^2+1=101$, $14^2+1=197$, $16^2+1=257$, $20^2+1=401$, $24^2+1=577$, $26^2+1=677$, $36^2+1=1297$, $40^2+1=1601$.

21 Twice a prime.
$3^2+1=2\cdot 5$, $5^2+1=2\cdot 13$, $9^2+1=2\cdot 41$, $11^2+1=2\cdot 61$, $15^2+1=2\cdot 113$, $19^2+1=2\cdot 181$, $25^2+1=2\cdot 313$, $29^2+1=2\cdot 421$, $35^2+1=2\cdot 613$, $39^2+1=2\cdot 761$, $45^2+1=2\cdot 1013$, $49^2+1=2\cdot 1201$.

22 $5^2+1=2\cdot 13\Rightarrow 6^2+4^2=2^2\cdot 13\Rightarrow 3^2+2^2=13$.
$9^2+1=2\cdot 41\Rightarrow 10^2+8^2=2^2\cdot 41\Rightarrow 5^2+4^2=41$.
$11^2+1=2\cdot 61\Rightarrow 12^2+10^2=2^2\cdot 61\Rightarrow 6^2+5^2=61$.
$15^2+1=2\cdot 113\Rightarrow 16^2+14^2=2^2\cdot 113\Rightarrow 8^2+7^2=113$.
$19^2+1=2\cdot 181\Rightarrow 20^2+18^2=2^2\cdot 181\Rightarrow 10^2+9^2=181$.
$25^2+1=2\cdot 313\Rightarrow 26^2+24^2=2^2\cdot 313\Rightarrow 13^2+12^2=313$.
$79^2+1=2\cdot 3121\Rightarrow 80^2+78^2=2^2\cdot 3121\Rightarrow 40^2+39^2=3121$.
$85^2+1=2\cdot 3613\Rightarrow 86^2+84^2=2^2\cdot 3613\Rightarrow 43^2+42^2=3613$.

23
$$\begin{aligned} 42^2+1=5\cdot 353 &\Rightarrow (42^2+1)(2^2+1)=5^2\cdot 353\\ &\Rightarrow (42\cdot 2+1\cdot 1)^2+(42\cdot 1-1\cdot 2)^2=5^2\cdot 353\\ &\Rightarrow 85^2+40^2=5^2\cdot 353\\ &\Rightarrow 17^2+8^2=353. \end{aligned}$$

$$\begin{aligned} 48^2+1=5\cdot 461 &\Rightarrow (48^2+1)(2^2+1)=5^2\cdot 461\\ &\Rightarrow (48\cdot 2-1\cdot 1)^2+(48\cdot 1+1\cdot 2)^2=5^2\cdot 461\\ &\Rightarrow 95^2+50^2=5^2\cdot 461\\ &\Rightarrow 19^2+10^2=461. \end{aligned}$$

$$52^2+1=5\cdot 541 \Rightarrow (52^2+1)(2^2+1)=5^2\cdot 541$$
$$\Rightarrow 105^2+50^2=5^2\cdot 541$$
$$\Rightarrow 21^2+10^2=541.$$

$$58^2+1=5\cdot 673 \Rightarrow (58^2+1)(2^2+1)=5^2\cdot 673$$
$$\Rightarrow 60^2+115^2=5^2\cdot 673$$
$$\Rightarrow 12^2+23^2=673.$$

24 $7^2+4^2=5\cdot 13 \Rightarrow (7^2+4^2)(2^2+1)=5^2\cdot 13$
$$\Rightarrow (7\cdot 2-4\cdot 1)^2+(7\cdot 1+4\cdot 2)^2=5^2\cdot 13$$
$$\Rightarrow 10^2+15^2=5^2\cdot 13$$
$$\Rightarrow 2^2+3^2=13.$$

$$9^2+2^2=5\cdot 17 \Rightarrow (9^2+2^2)(2^2+1)=5^2\cdot 17$$
$$\Rightarrow (9\cdot 2+2\cdot 1)^2+(9\cdot 1-2\cdot 2)^2=5^2\cdot 17$$
$$\Rightarrow 20^2+5^2=5^2\cdot 17$$
$$\Rightarrow 4^2+1^2=17.$$

25 $33^2+1=2\cdot 5\cdot 109 \Rightarrow 34^2+32^2=2^2\cdot 5\cdot 109$
$$\Rightarrow 17^2+16^2=5\cdot 109$$
$$\Rightarrow (17^2+16^2)(2^2+1)=5^2\cdot 109$$
$$\Rightarrow (17\cdot 2+16\cdot 1)^2+(17\cdot 1-16\cdot 2)^2=5^2\cdot 109$$
$$\Rightarrow 50^2+15^2=5^2\cdot 109$$
$$\Rightarrow 10^2+3^2=109.$$

$$67^2+1=2\cdot 5\cdot 449 \Rightarrow 68^2+66^2=2^2\cdot 5\cdot 449$$
$$\Rightarrow 34^2+33^2=5\cdot 449$$
$$\Rightarrow (34^2+33^2)(2^2+1)=5^2\cdot 449$$
$$\Rightarrow (34\cdot 2-33\cdot 1)^2+(34\cdot 1+33\cdot 2)^2=5^2\cdot 449$$
$$\Rightarrow 35^2+100^2=5^2\cdot 449$$
$$\Rightarrow 7^2+20^2=449.$$

26 $11^2+13^2=2\cdot 5\cdot 29 \Rightarrow 24^2+2^2=2^2\cdot 5\cdot 29$
$$\Rightarrow 12^2+1^2=5\cdot 29$$
$$\Rightarrow (12^2+1^2)(2^2+1^2)=5^2\cdot 29$$
$$\Rightarrow (12\cdot 2+1\cdot 1)^2+(12\cdot 1-2\cdot 1)^2=5^2\cdot 29$$
$$\Rightarrow 25^2+10^2=5^2\cdot 29$$
$$\Rightarrow 5^2+2^2=29.$$

27 Since $x^2+y^2=13$, $x, y=\pm 2, \pm 3$.
In fact $(x, y)=(2, -3), (-2, 3), (3, 2)$ or $(-3, -2)$.

$$(34^2+1)(2^2+3^2)=13^2\cdot 89$$
$$\Rightarrow (34\cdot 2-1\cdot 3)^2+(34\cdot 3+1\cdot 2)^2=13^2\cdot 89$$
$$\Rightarrow 65^2+104^2=13^2\cdot 89$$
$$\Rightarrow 5^2+8^2=89.$$

28 $n^2 \equiv -1$ and $x^2 \equiv -y^2 \Rightarrow n^2x^2 \equiv y^2$ and so $nx \equiv \pm y \pmod p$. If $nx \equiv y$ then $ny \equiv -x$ and $p | nx - y$, $ny + x$, so

$$\left(\frac{nx-y}{p}\right)^2 + \left(\frac{ny+x}{p}\right)^2 = k.$$

The argument is similar if $nx \equiv -y \pmod p$.

29 Since $x^2 + y^2 = 29$, $y = \pm 2, \pm 5$.
$(x, y) = (2, -5), (-2, 5), (5, 2)$ or $(-5, -2)$.

$$(113^2+22^2)(2^2+52) = 29^2 \cdot 457$$
$$\Rightarrow (113 \cdot 2 - 22 \cdot 5)^2 + (113 \cdot 5 + 22 \cdot 2)^2 = 29^2 \cdot 457$$
$$\Rightarrow 116^2 + 609^2 = 29^2 \cdot 457$$
$$\Rightarrow 4^2 + 21^2 = 457.$$

30 $m^2 \equiv -n^2$ and $x^2 \equiv -y^2 \Rightarrow m^2x^2 \equiv n^2y^2$ and so $mx \equiv \pm ny \pmod p$. If $mx \equiv ny$, $mny \equiv m^2x \equiv -n^2x$, so $my \equiv -nx \pmod p$ and $p | mx - ny$, $my + nx$, whence

$$\left(\frac{mx-ny}{p}\right)^2 + \left(\frac{my+nx}{p}\right)^2 = k.$$

The argument is similar if $mx \equiv -ny \pmod p$.

31 If $p, q | m^2+1$ and $p, q \geqslant m+1$, $pq \geqslant m^2+2m+1 > m^2+1$, so m^2+1 has at most one prime factor $> m$.

32 If $p | m^2+1$ and $p < m$, then $p | (m-p)^2+1$ and p is the sum of two squares by hypothesis.
Suppose $m^2+1 = p_1p_2 \ldots p_n$, where the p_i are primes, not necessarily distinct. By q 31, at most one of the $p_i > m$. Suppose $p_n > m$. All other prime factors are equal to the sum of two squares by hypothesis, so by repeated use of q 30,

$$\frac{m^2+1}{p_1}, \frac{m^2+1}{p_1p_2}, \ldots, \frac{m^2+1}{p_1p_2 \ldots p_{n-1}}$$

are each equal to the sum of two squares. This last number is equal to p_n.

33 Since $2 = 1^2 + 1^2$ is the sum of two squares, this forms a basis for an induction on m. The inductive step was established in q 32, so every prime dividing a number of the form n^2+1 is expressible as the sum of two squares. By q 19, these are precisely the primes congruent to 1 (mod 4) and the prime 2.

34 By q 33, the prime 2 and every prime congruent to 1 (mod 4) can be expressed as the sum of two squares, so by q 15, these primes can occur to any power in the factorisation of the sum of two squares. By q 10, a prime congruent to 3 (mod 4) can only occur to an even power.

35 No.

36 $n^2 \equiv -1$, $a^2 \equiv -b^2$, $c^2 \equiv -d^2 \Rightarrow a^2n^2 \equiv b^2$ and $c^2n^2 \equiv d^2 \pmod p$. So $an \equiv \pm b$, $cn \equiv \pm d \pmod p$.
If $an \equiv b$ and $cn \equiv d$ or if $an \equiv -b$ and $cn \equiv -d \pmod p$, then $ac + bd \equiv ac(1+n^2) \equiv 0$ and $ad - bc \equiv \pm ac(n-n) \equiv 0 \pmod p$. So the equation

$$p^2 = (a^2+b^2)(c^2+d^2) = (ac+bd)^2 + (ad-bc)^2$$

can be divided through by p^2. This gives an expression for 1 as the sum of two squares, one of which must be 0. If $ad = bc$,

$$d^2(a^2+b^2) = b^2c^2 + d^2b^2 = b^2(c^2+d^2),$$

so $b^2 = d^2$. Similar arguments pertain with the alternative signs. A simpler proof of the uniqueness of the set $\{a^2, b^2\}$ follows from the uniqueness of factorisation in the Gaussian integers.

What this question establishes is that if a prime number can be expressed as the sum of two squares, then this can be done in only one way.

37 One square: 1, 4, 9, 16.
Two squares: 2, 5, 8, 10, 13, 17, 18, 20.
Three squares: 3, 6, 11, 12, 14, 19.
Four squares: 7, 15.

38
$$\begin{aligned}(a+b\mathrm{i}+c\mathrm{j}+d\mathrm{k})(a-b\mathrm{i}-c\mathrm{j}-d\mathrm{k}) &= a^2 - ab\mathrm{i} - ac\mathrm{j} - ad\mathrm{k}\\ &\quad + ab\mathrm{i} + b^2 - bc\mathrm{k} + bd\mathrm{j}\\ &\quad + ac\mathrm{j} + bc\mathrm{k} + c^2 - cd\mathrm{i}\\ &\quad + ad\mathrm{k} - bd\mathrm{j} + cd\mathrm{i} + d^2 = a^2+b^2+c^2+d^2.\end{aligned}$$

$$\begin{aligned}\alpha\beta &= (a+b\mathrm{i}+c\mathrm{j}+d\mathrm{k})(x+y\mathrm{i}+z\mathrm{j}+t\mathrm{k})\\ &= ax + ay\mathrm{i} + az\mathrm{j} + at\mathrm{k}\\ &\quad + bx\mathrm{i} - by + bz\mathrm{k} - bt\mathrm{j}\\ &\quad + cx\mathrm{j} - cy\mathrm{k} - cz + ct\mathrm{i}\\ &\quad + dx\mathrm{k} + dy\mathrm{j} - dz\mathrm{i} - dt\\ &= (ax - by - cz - dt)\\ &\quad + \mathrm{i}(ay + bx + ct - dz)\\ &\quad + \mathrm{j}(az - bt + cx + dy)\\ &\quad + \mathrm{k}(at + bz - cy + dx)\end{aligned}$$

So

$$\begin{aligned}\overline{\beta\alpha} &= (ax - by - cz - dt)\\ &\quad - \mathrm{i}(xb + ya + zd - tc)\\ &\quad - \mathrm{j}(xc - yd + za + tb)\\ &\quad - \mathrm{k}(xd + yc - zb + ta) = \bar\alpha\bar\beta\end{aligned}$$

39 With α and β as in q 38,
$(a^2+b^2+c^2+d^2)(x^2+y^2+z^2+t^2) = \alpha\bar\alpha\beta\bar\beta$.

Now $\beta\bar{\beta}$ is real and so commutes with $\bar{\alpha}$. Thus

$$\begin{aligned}\alpha\bar{\alpha}\beta\bar{\beta} &= \alpha\beta\bar{\beta}\bar{\alpha} = \alpha\beta\overline{\alpha\beta}\\ &= (ax-by-cz-dt)^2+(ay+bx+ct-dz)^2\\ &\quad +(az-bt+cx+dy)^2+(at+bz-cy+dx)^2\\ &= \text{sum of four squares.}\end{aligned}$$

40 If every prime congruent to 3 (mod 4) can be expressed as the sum of four squares, then the theorem is established.

41 There are $\frac{1}{2}(p-1)$ quadratic residues in $\mathbf{Z}_p$. 0 is also a square. So $\mathbf{Z}_p$ contains $\frac{1}{2}(p+1)$ squares in all.
The two sets each contain $\frac{1}{2}(p+1)$ elements in $\mathbf{Z}_p$, and $\frac{1}{2}(p+1)+\frac{1}{2}(p+1)=p+1>$ number of distinct elements in $\mathbf{Z}_p$.

42 $x^2\equiv(p-x)^2 \pmod p$, so if $0\leqslant x<p$, either x or $p-x<\frac{1}{2}p$, and there exist integers x, y, with $0\leqslant x$, $y<\frac{1}{2}p$ such that $x^2+y^2+1^2+0^2\equiv 0 \pmod p$ and $x^2+y^2+1^2+0^2=mp$. Now $x^2, y^2<(\frac{1}{2}p)^2$, so $x^2+y^2+1<\frac{1}{2}p^2+1<p^2$ for $p>2$, and so $m<p$.
An alternative proof of the essential result here is available using q 3.82.

43 Either 0, 2 or 4 are odd. $a^2\equiv a \pmod 2$.

44 $a+b$ and $a-b$ are both even.

45 $k=\frac{1}{2}$.

46 $1^2+3^2+5^2+7^2=84$,
so $2^2+1^2+6^2+1^2=\frac{1}{2}\cdot 84=42$, using q 45,
so $1^2+1^2+2^2+6^2=42$,
and $1^2+0^2+4^2+2^2=\frac{1}{2}\cdot 42$, using q 45,
so $0^2+1^2+2^2+4^2=21$.

47 If mp is even, a, b, c, d consist of pairs with the same parity. If a and b, and c and d have the same parity we can use q 45 to express $\frac{1}{2}mp$ as the sum of four squares.

48 $1^2+2^2+3^2+5^2\equiv 0 \pmod 3$, so $1x+2y+3z+5t\equiv 0 \pmod 3$ and $x^2+y^2+z^2+t^2\equiv 0 \pmod 3$.
$1y\equiv 2x$ and $3t\equiv 5z \pmod 3$ establish divisibility by 3 for the third expression.
$1z\equiv 3x$ and $2t\equiv 5y \pmod 3$ give the result for the fourth.
$1t\equiv 5x$ and $2z\equiv 3y \pmod 3$ give the result for the fifth.
Choose $x=1$, $y=-1$, $z=0$, $t=-1$, then

$$\begin{aligned}&(1^2+2^2+3^2+5^2)(1^2+(-1)^2+0^2+(-1)^2)\\ &\quad=(1-2+0-5)^2+(-1-2+3+0)^2+(0-2-3+5)^2\\ &\qquad+(-1-0-3-5)^2,\end{aligned}$$

using the identity of q 39, with the signs of b, c, and d changed, so $3^2\cdot 13=6^2+0^2+0^2+9^2$, and $13=2^2+0^2+0^2+3^2$.

49 Choose $x=1$, $y=1$, $z=2$, $t=2$.

$$(1^2+1^2+2^2+17^2)(1^2+1^2+2^2+2^2)$$
$$=(1+1+4+34)^2+(1-1-4+34)^2$$
$$+(2+2-2-17)^2+(2-2+2-17)^2$$

so $\quad 2\cdot 5^2\cdot 59=40^2+30^2+15^2+15^2$

and $\quad 2\cdot 59=8^2+6^2+3^2+3^2$

so $\quad 59=7^2+1^2+3^2+0^2$.

50 Since $a^2+b^2+c^2+d^2\equiv 0 \pmod m$,

$ax+by+cz+dt\equiv 0 \pmod m$

and $\quad x^2+y^2+z^2+t^2\equiv 0 \pmod m$.

Now $ay\equiv bx$ and $ct\equiv dz \pmod m$ establish divisibility by m for the third expression,

$az\equiv cx$ and $bt\equiv dy \pmod m$ for the fourth,

and

$at\equiv dx$ and $bz\equiv cy \pmod m$ for the fifth.

The whole equation

$$(a^2+b^2+c^2+d^2)(x^2+y^2+z^2+t^2)$$
$$=(ax+by+cz+dt)^2+(ay-bx-ct+dz)^2$$
$$+(az+bt-cx-dy)^2+(at-bz+cy-dx)^2$$

obtained from the identity of q 39 by changing the signs of b, c and d, can be divided through by m^2.

If numerically least residues have been chosen for x, y, z and t, then x^2, y^2, z^2, $t^2<(\frac{1}{2}m)^2$, so

$x^2+y^2+z^2+t^2=m'm<m^2$.

Dividing the previous equation by m^2 gives $m'p$ as the sum of four squares.

51 If p is an odd prime, then a multiple of p less than p^2 can be expressed as the sum of four squares by q 42. If this multiple is greater than p, then by q 45 for an even multiple and by q 50 for an odd multiple, a lesser multiple of p is expressible as the sum of four squares.

By repeating the procedure of q 45 and/or q 50 a sufficient number of times, p can be expressed as the sum of four squares.

This is an example of Fermat's method of descent.

52 With the prime number $2=1^2+1^2$, and q 51, every prime number can be expressed as the sum of four squares. By q 39, the set of integers expressible as the sum of four squares is closed under multiplication, so every positive integer is thus expressible.

53 Only 7 and 15 cannot be written in this way.

54 $0^2 \equiv 0,\ 1^2 \equiv 1,\ 2^2 \equiv 4,\ 3^2 \equiv 1,\ 4^2 \equiv 0,\ 5^2 \equiv 1,\ 6^2 \equiv 4,\ 7^2 \equiv 1 \pmod 8$.

				+1			+4	
0	1	4	1	2	5	4	5	0
1	2	5	2	3	6	5	6	1
4	5	0	5	6	1	0	1	4

Among the 27 possibilities, 7 does not appear among the sums of three squares modulo 8.

$0^2 \equiv 0,\ 1^2 \equiv 1,\ 2^2 \equiv 0,\ 3^2 \equiv 1 \pmod 4$

		+1	
0	1	1	2
1	2	2	3

Thus $x^2+y^2+z^2 \equiv 0 \pmod 4$ implies $x^2 \equiv y^2 \equiv z^2 \equiv 0 \pmod 4$. If $4m = x^2+y^2+z^2$, x, y, and z are even, so the equation can be divided through by 4.

Thus if $4^h(8k+7)$ were expressible as the sum of three squares, so would $4^{h-1}(8k+7)$, $4^{h-2}(8k+7), \ldots, 8k+7$ be expressible as the sum of three squares. This is false for $8k+7$, so it is false for its predecessors.

Historical note

Through his study of Pythagorean triangles, Diophantus (c. A.D. 200) knew how to express $(a^2+b^2)(c^2+d^2)$ as the sum of two squares. The same expression was derived by L. Euler in 1770 using complex numbers. Diophantus also knew that a sum of two squares cannot be congruent to 3 (mod 4). Fermat used his method of descent to establish that a prime number congruent to 1 (mod 4) can be expressed as the sum of two squares, and this leads to a complete account of the sums of two squares. Fermat also showed that the expression for a prime as the sum of two squares is unique (q 6.36).

Bachet (in 1621) noticed that Diophantus assumed that every number is either a square or the sum of two, three or four squares. Fermat was the first to claim a proof. In 1748, L. Euler showed how to express

$$(a^2+b^2+c^2+d^2)(x^2+y^2+z^2+t^2)$$

as a sum of four squares, and in 1751 he proved that every prime was a factor of a number of the form $1+a^2+b^2$. Using Euler's work, J. L. Lagrange finally published a complete proof in 1770. Quaternions were constructed by W. R. Hamilton in 1843.

Fermat identified those numbers which are not expressible as the sum of three squares (q 6.54) and in 1798 A. M. Legendre proved that these were the only such numbers.

7

Partitions

Ferrers' graphs

1 $1+1+1+1=2+1+1=2+2=3+1=4$.
Because of this, the positive integer 4 is said to have five *partitions.* Exhibit the seven partitions of 5 and the eleven partitions of 6. How many partitions of 7 are there?

2 The five partitions of 4 may be exhibited thus:

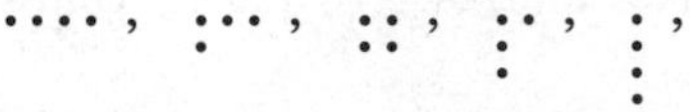

and these are called the *graphs* of the partitions. Exhibit the graphs of the partitions of 5 and of 6.

3 The columns of the graph :·· give the partition $2+1+1$. The rows of the same graph give the partition $3+1$. Interchanging rows and columns converts :·· to ⋮·. Such pairs of partitions are said to be *conjugate.* Which partition of 4 is not conjugate to any other partition of 4? Suggest a name for such a partition. Find all the partitions of numbers up to 10 which are not conjugate to any other partition.

4 Exhibit all the partitions of the numbers up to 10 into distinct odd parts.

5 Does every odd number have at least one self-conjugate partition?

6 If a number has a self-conjugate partition, must it have at least one partition into distinct odd parts?

7 If a number has a partition into distinct odd parts, must it have a self-conjugate partition?

8 Conjecture and prove a theorem about the numbers of self-conjugate partitions of a number n and the number of partitions of n into distinct odd parts.

9 Exhibit the partitions of the numbers from 1 to 7 which have only the numbers 1 and/or 2 as parts.

10 How many partitions of $2n$ are there which have only the numbers 1 and/or 2 as parts?
How many partitions of $2n+1$ are there which have only the numbers 1 and/or 2 as parts?
The number of partitions of n in which each part is 1 or 2 is denoted by $p_2(n)$.

11 Exhibit the partitions of the numbers from 1 to 7 which have only one or two not necessarily distinct parts.

12 How many partitions of $2n$ are there which have only one or two not necessarily distinct parts?
What is the number of partitions of $2n+1$ which have only one or two not necessarily distinct parts?

13 If a partition has only the numbers 1 and/or 2 as parts, what can be said about its graph?

14 If a partition has only one or two parts, what can be said about its graph?

15 Use an argument about graphs to link the number of partitions having only the numbers 1 and/or 2 as parts with the number of partitions having only one or two parts.

Generating functions

16 Calculate the first eight terms in the expansion of

$$(1+x+x^2+\ldots+x^n+\ldots)(1+x^2+x^4+\ldots+x^{2n}+\ldots).$$

For these eight terms, compare the coefficients of x^n with the number of partitions of n having only the numbers 1 and/or 2 as parts.
By writing the exponents in the first bracket as sums of 1s, and the exponents in the second bracket as sums of 2s, establish generally that

$$(1+x+x^2+\ldots)(1+x^2+x^4+\ldots)=1+\sum_{n=1}^{\infty} p_2(n)x^n,$$

[[139]]

Table 7.1. *Ferrers' graphs*

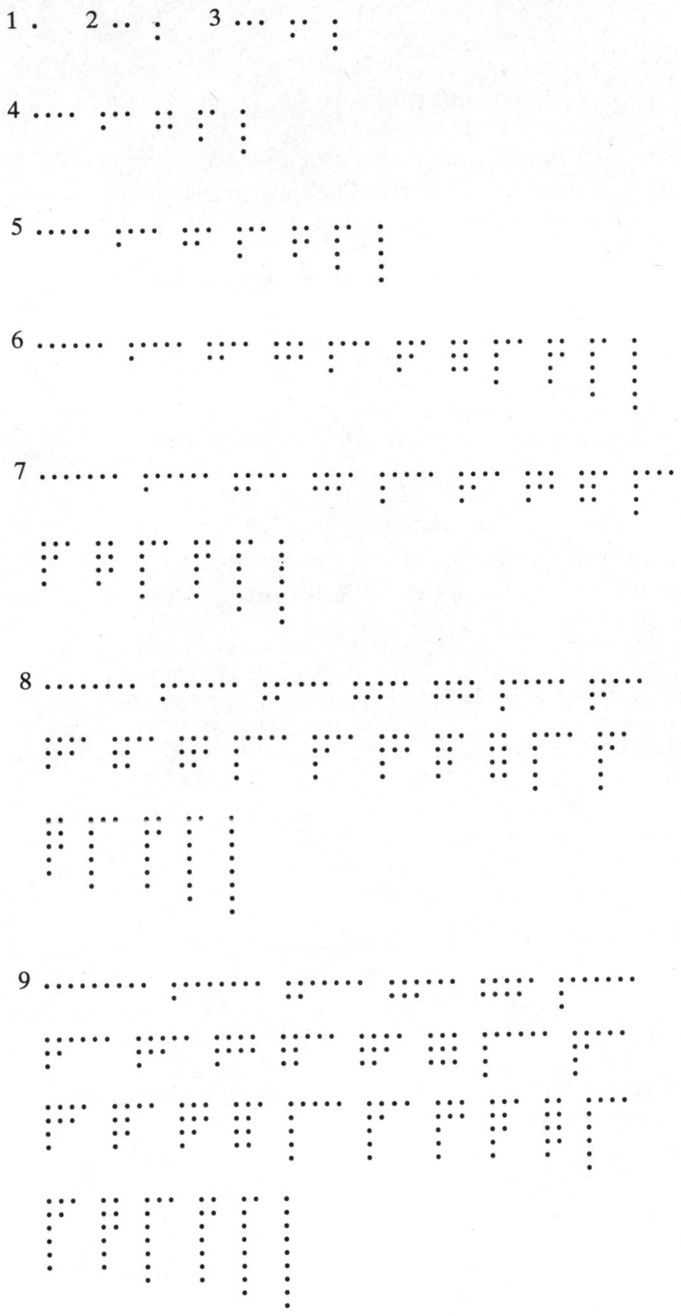

or, using the formula for a geometric progression when $|x|<1$, that

$$\frac{1}{(1-x)(1-x^2)}=1+\sum_{n=1}^{\infty} p_2(n)x^n.$$

The function $p_2(n)$ is defined in q 10.

17 Count the partitions of the numbers 1 to 9 which have only the numbers 1, 2 and/or 3 as parts. The nine answers are denoted by $p_3(1)$, $p_3(2)$, $p_3(3)$, $p_3(4)$, $p_3(5)$, $p_3(6)$, $p_3(7)$, $p_3(8)$ and $p_3(9)$.

18 Count the partitions of the numbers 1 to 9 into at most three parts.

19 How can the graphs of the partitions of q 17 and q 18 respectively be described?

20 Evaluate the first seven terms in the expansion of

$$(1+x+x^2+\ldots+x^n+\ldots)(1+x^2+x^4+\ldots+x^{2n}+\ldots)$$
$$\times(1+x^3+x^6+\ldots+x^{3n}+\ldots).$$

For these seven terms, compare the coefficients of x^n with $p_3(n)$. Explain why the coefficient of x^n in this expansion is $p_3(n)$ and deduce that, when $|x|<1$,

$$\frac{1}{(1-x)(1-x^2)(1-x^3)}=1+\sum_{n=1}^{\infty} p_3(n)x^n.$$

21 Presuming that $|x|<1$, make a conjecture about the coefficients a_n in the expansion

$$\frac{1}{(1-x)(1-x^2)(1-x^3)(1-x^4)}=1+\sum_{n=1}^{\infty} a_nx^n.$$

Justify or revise your conjecture.
Conventionally, a_n is written $p_4(n)$.

22 The number of partitions of n into parts $\leqslant m$ is denoted by $p_m(n)$. Suggest a generating function for $p_m(n)$ by analogy with q 16, q 20 and q 21.

23 If $p(n)$ is the total number of partitions of n, what comparison may be made between $p_m(n)$ and $p(n)$

(i) when $m<n$,
(ii) when $m=n$,
(iii) when $m>n$?

24 Suggest an infinite product which is a generating function for $p(n)$.

25 For what number k is $p_m(k)$ equal to the number of partitions of n containing m as largest part?

26 For what number l is $p_l(n)$ equal to the number of partitions of n with every part $<m$?

27 Use q 25 and q 26 to prove that $p_m(n)=p_m(n-m)+p_{m-1}(n)$ when $n>m$.
What value must be given to $p_m(0)$ to make the equation valid when $n=m$?
What value must be given to $p_m(n-m)$ to make the equation valid when $n<m$?

28 Use q 27 to prove that if

$$F_m(x)=1+\sum_{n=1}^{\infty} p_m(n)x^n$$

then

$$F_m(x)=x^m F_m(x)+F_{m-1}(x).$$

Deduce the generating function you suggested in q 22 by induction.

29 If

$$\frac{1}{(1-x)(1-x^3)(1-x^5)\dots(1-x^{2n-1})\dots}=1+\sum_{n=1}^{\infty} q_n x^n,$$

what kind of partitions of n are counted by q_n?

30 If

$$\frac{1}{(1-x^2)(1-x^4)(1-x^6)\dots(1-x^{2n})\dots}=1+\sum_{n=1}^{\infty} r_n x^n,$$

what kind of partitions of n are counted by r_n?

31 If

$$(1+x)(1+x^2)(1+x^3)\dots(1+x^n)\dots=1+\sum_{n=1}^{\infty} s_n x^n,$$

what kind of partitions of n are counted by s_n?

32 If

$$(1+x)(1+x^3)(1+x^5)\dots(1+x^{2n-1})\dots=1+\sum_{n=1}^{\infty} t_n x^n,$$

what kind of partitions of n are counted by t_n?

⟦141⟧

33 What can you deduce about the partitions of n from the identity

$$(1+x)(1+x^2)(1+x^3)\dots(1+x^n)\dots$$
$$=\frac{1-x^2}{1-x}\frac{1-x^4}{1-x^2}\frac{1-x^6}{1-x^3}\dots\frac{1-x^{2n}}{1-x^n}\dots$$
$$=\frac{1}{1-x}\frac{1}{1-x^3}\frac{1}{1-x^5}\dots\frac{1}{1-x^n}\dots?$$

34 If

$$(1+x+x^2)(1+x^2+x^4)(1+x^3+x^6)\dots(1+x^n+x^{2n})\dots$$
$$=1+\sum_{n=1}^{\infty}u_n x^n,$$

what kind of partitions of n are counted by u_n?

35 Prove that the number of partitions of n with no part repeated more than twice is equal to the number of partitions of n with no part divisible by 3.

Euler's theorem

36 For $n=1,\dots,9$, find $E(n)$, the number of partitions of n into an even number of unequal parts, and $O(n)$, the number of partitions of n into an odd number of unequal parts. For which n, up to 9, does $E(n)=O(n)$?

37 Exhibit the graphs of some partitions into unequal parts. What characterises these graphs?

38 For each of the partitions that you have illustrated in q 37, determine whether another partition into unequal parts is obtained by removing the smallest part (k, say) and adding 1 each to the k largest parts. Under what circumstances would this transformation not give rise to the graph of a new partition into unequal parts? If the transformation is possible, we call this transformation α.

39 For each of the partitions that you have illustrated in q 37, determine whether another partition into unequal parts is obtained by removing the nodes on a north-east south-west line through the last node of the largest part and inserting these nodes as a new smallest part.
Under what circumstances would this transformation not give rise to the graph of a new partition into unequal parts?
If the transformation is possible, how does it affect the number of parts? We call this transformation β.

⟦141⟧

40 Exhibit two more graphs of partitions into unequal parts, distinct from those you have shown in q 37. For each of these, let k be the smallest part and let l be the number of nodes on a north-east line through the last node of the largest part.
If $k \leqslant l$, perform α as in q 38 if possible.
If $k > l$, perform β as in q 39 if possible.

41 If, in q 40, $k \leqslant l$ and α is not possible, deduce that $k = l$ and determine how many parts there are in the partition. Prove that the partition in question is

$$(2l-1)+(2l-2)+\ldots+(l+1)+l = l^2+\tfrac{1}{2}(l-1)l$$
$$= \tfrac{1}{2}l(3l-1).$$

42 If, in q 40, $k > l$ and β is not possible, deduce that $k = l+1$ and determine how many parts there are in the partition.
Prove that the partition in question is

$$2l+(2l-1)+\ldots+(l+2)+(l+1) = l^2+\tfrac{1}{2}l(l+1)$$
$$= \tfrac{1}{2}l(3l+1).$$

43 If $n \neq \frac{1}{2}l(3l \pm 1)$, prove that a transformation T of the unequal partitions of n, defined by $T = \alpha$ for those partitions for which $k \leqslant l$, and by $T = \beta$ for those partitions for which $k > l$, is one–one and maps every even partition onto an odd partition and vice-versa. Deduce that $E(n) = O(n)$.

44 If $n = \frac{1}{2}l(3l \pm 1)$, show that the only partition of n into unequal parts which cannot be paired with one of opposite parity by the procedure of q 40 has l parts. Deduce the values of $E(n) - O(n)$ for every n.

45 Calculate the first nine terms in the expansion of

$$(1-x)(1-x^2)(1-x^3)\ldots(1-x^n)\ldots$$

46 Establish that, if $|x| < 1$,

$$(1-x)(1-x^2)(1-x^3)\ldots(1-x^n)\ldots = 1+\sum_{n=1}^{\infty}(E(n)-O(n))x^n.$$

47 Prove that, if $|x| < 1$,

$$(1-x)(1-x^2)(1-x^3)\ldots(1-x^n)\ldots = \sum_{n=-\infty}^{\infty}(-1)^n x^{\frac{1}{2}n(3n-1)}.$$

(*Euler's theorem*)

Notes and answers

This chapter is by way of an interlude. It does not depend on the theorems of the preceding chapters, nor do the following chapters depend on what is here. For concurrent reading see bibliography: Andrews (1971), Chrystal (1964).

1 $2+1+1$ gives the *same* partition of 4 as $1+2+1$ and $1+1+2$. We do not distinguish between the different orders in which the sum may occur, we only study the parts themselves.

$$1+1+1+1+1=2+1+1+1=2+2+1=3+1+1=3+2=4+1=5$$

$$1+1+1+1+1+1=2+1+1+1+1=2+2+1+1=2+2+2=3+1+1+1$$
$$=3+2+1=3+3=4+1+1=4+2=5+1=6$$

There are fifteen partitions of 7.
The number of partitions of n is denoted by $p(n)$, so $p(4)=5$, $p(5)=7$, $p(6)=11$, $p(7)=15$.

2 Most books exhibit these arrays of nodes summing downwards rather than across. The graph ..... then represents the partition of 5 into one part, not into five parts as we have.

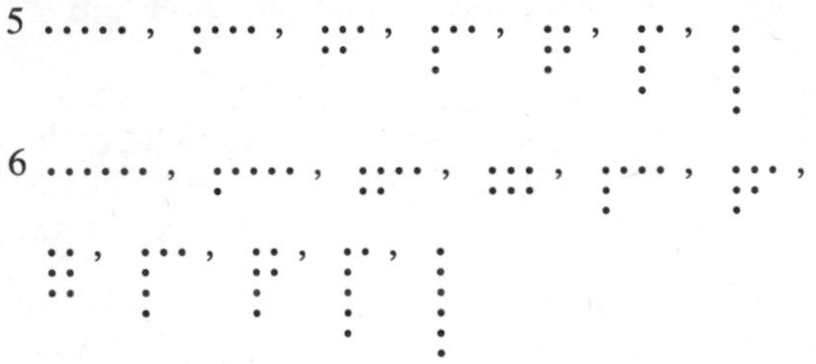

These graphs are properly called Ferrers' graphs and this distinguishes them from those graphs which have edges and regions.

3 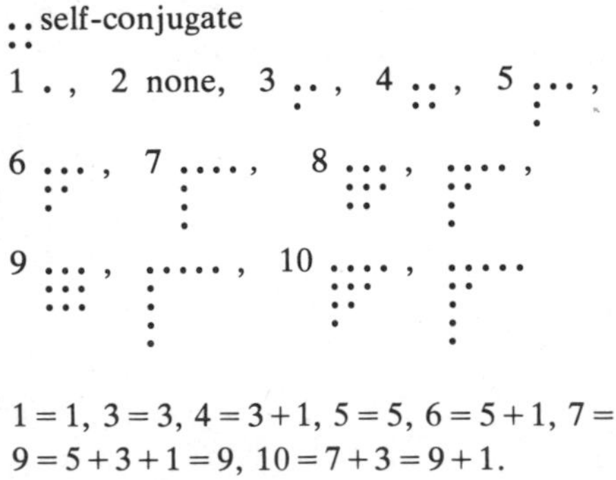

4 $1=1$, $3=3$, $4=3+1$, $5=5$, $6=5+1$, $7=7$, $8=5+3=7+1$, $9=5+3+1=9$, $10=7+3=9+1$.

5 $2n+1$ may be partitioned into $(n+1)+\underbrace{1+1+1\ldots+1}_{n\text{ times}}$

for which the graph is

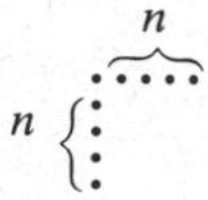

6 Yes, for example:

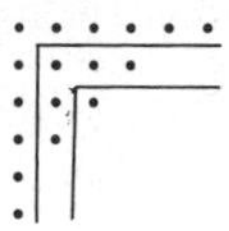

11
+5
+1

7 The odd numbers each give L-shaped arrays as in q 5 and if they are distinct and are fitted together in order of size as in q 6, the resulting array is the graph of a partition.
Since each part added is self-conjugate and is added to a self-conjugate graph, the result is a self-conjugate graph.

9 $1=1$
$2=2=1+1$
$3=2+1=1+1+1$
$4=2+2=2+1+1=1+1+1+1$
$5=2+2+1=2+1+1+1=1+1+1+1+1$
$6=2+2+2=2+2+1+1=2+1+1+1+1=1+1+1+1+1+1$
$7=2+2+2+1=2+2+1+1+1=2+1+1+1+1+1$
$=1+1+1+1+1+1+1$

10 $2n=n$ twos $=(n-1)$ twos $+2$ ones $=(n-k)$ twos $+2k$ ones, $k\leqslant n$. Thus there are exactly $n+1$ partitions of this type. $2n+1=(2n)+1$, so $2n+1$ also has exactly $n+1$ partitions of this type.
$p_2(n)=[\frac{1}{2}n]+1$.

11 1, $2=1+1$, $3=2+1$, $4=3+1=2+2$, $5=4+1=3+2$, $6=5+1=4+2=3+3$, $7=6+1=5+2=4+3$.

12 $n+1$, $n+1$.

13 At most two rows.

14 At most two columns.

15 These two types occur in conjugate pairs, so $p_2(n)$, the number of partitions of n with each part $\leqslant 2$ is equal to the number of partitions of n into at most two parts.

16 $1+x+2x^2+2x^3+3x^4+3x^5+4x^6+4x^7+\ldots$

Each partition of n into 1s and/or 2s adds one to the coefficient of x^n in the product.

If $n=\underbrace{2+2+\ldots+2}_{a\text{ times}}+\underbrace{1+1+\ldots+1}_{b\text{ times}}$, then $x^n=x^b x^{2a}$.

$\dfrac{1}{(1-x)(1-x^2)}$ is called the *generating function* for $p_2(n)$.

17 $p_3(1)=1$, $p_3(2)=2$, $p_3(3)=3$, $p_3(4)=4$, $p_3(5)=5$, $p_3(6)=7$, $p_3(7)=8$, $p_3(8)=10$, $p_3(9)=12$.

18 Same answers as in q 17.

19 The graphs for q 17 have one, two or three rows.
The graphs for q 18 have one, two or three columns.
The partitions of these two types are paired by conjugacy.

20 $1+x+2x^2+3x^3+4x^4+5x^5+7x^6+8x^7+10x^8+12x^9+\ldots$

If $n=\underbrace{3+3+\ldots+3}_{a\text{ times}}+\underbrace{2+2+\ldots+2}_{b\text{ times}}+\underbrace{1+1+\ldots+1}_{c\text{ times}}$

then $x^n=x^c x^{2b} x^{3a}$ and so this partition adds one to the coefficient of x^n in the product. Every unit in the coefficient of x^n is obtained in this way from a partition of this type.

When $|x|<1$, $1+x+x^2+\ldots=\dfrac{1}{1-x}$

$$1+x^2+x^4+\ldots=\frac{1}{1-x^2}$$

$$1+x^3+x^6+\ldots=\frac{1}{1-x^3}$$

$\dfrac{1}{(1-x)(1-x^2)(1-x^3)}$ is the generating function for $p_3(n)$.

21 The left hand side is

$(1+x+x^2+\ldots)(1+x^2+x^4+\ldots)(1+x^3+x^6+\ldots)(1+x^4+x^8+\ldots)$,

so each product x^n comes from a partition of n into parts $\leqslant 4$, and the coefficient a_n counts the number of such partitions.

22 $\dfrac{1}{(1-x)(1-x^2)(1-x^3)\ldots(1-x^m)}$

is the obvious generalisation of the argument in n 21. Proof follows in q 28.

23 When $m<n$, $p(n)>p_m(n)$.
When $m\geqslant n$, $p(n)=p_m(n)$.

24 $\dfrac{1}{(1-x)(1-x^2)(1-x^3)\ldots(1-x^m)\ldots}$

25 $k=n-m$.

26 $l=m-1$.

27 $p_m(n)=$ number of partitions of n with parts $\leqslant m$,
$=$ number of partitions of n with greatest part m + number of partitions of n with parts $\leqslant m-1$,
$=p_m(n-m)+p_{m-1}(n)$.
Take $p_m(0)=1$, for when $m=n$, there is just one partition of n with greatest part $=m$.
Take $p_m(n-m)=0$ when $m>n$, for then $p_m(n)=p_{m-1}(n)$.

28 $$F_m(x)=1+\sum_{n=1}^{\infty} p_m(n)x^n$$
$$=1+\sum_{n=1}^{\infty}(p_m(n-m)+p_{m-1}(n))x^n$$
$$=1+\sum_{n=1}^{\infty} p_{m-1}(n)x^n+\sum_{n=1}^{\infty} p_m(n-m)x^{n-m}x^m.$$

Now adopting the values for $p_m(n-m)$ in q 27 when $n\leqslant m$, we have

$$F_m(x)=F_{m-1}(x)+\left[1+\sum_{n-m=1}^{\infty} p_m(n-m)x^{n-m}\right]x^m$$
$$=F_{m-1}(x)+x^mF_m(x).$$

29 q_n is the number of partitions of n into odd parts.

30 r_n is the number of partitions of n into even parts.

31 s_n is the number of partitions of n into distinct parts.

32 t_n is the number of partitions of n into distinct odd parts.

33 The number of partitions of n into distinct parts is equal to the number of partitions of n into odd parts.

34 Each element of the product comes from multiplying together one term from each bracket, either x^0 or x^n or x^{n+n} from $1+x^n+x^{2n}$, so in the corresponding partition no part occurs more than twice.

35 $$(1+x+x^2)(1+x^2+x^4)\ldots(1+x^n+x^{2n})\ldots$$
$$=\frac{1-x^3}{1-x}\cdot\frac{1-x^6}{1-x^2}\cdots\frac{1-x^{3n}}{1-x^n}\cdots$$
$$=\frac{1}{(1-x)(1-x^2)(1-x^4)(1-x^5)\ldots(1-x^{3n-2})(1-x^{3n-1})\ldots}$$

36

$E(1)=0$	$O(1)=1$	$E(n)-O(n)=-1$
$E(2)=0$	$O(2)=1$	-1
$E(3)=1$	$O(3)=1$	0
$E(4)=1$	$O(4)=1$	0
$E(5)=2$	$O(5)=1$	1
$E(6)=2$	$O(6)=2$	0
$E(7)=3$	$O(7)=2$	1
$E(8)=3$	$O(8)=3$	0
$E(9)=4$	$O(9)=4$	0

37 Lines joining last nodes of adjacent parts have gradient $\geqslant 1$.

38 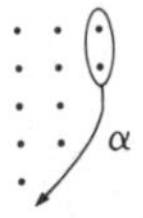

possible.  not possible. 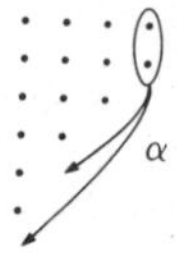

possible.

α is possible when the number of parts > smallest part.

α reduces the number of parts by one, and so changes the parity of the partition.

39 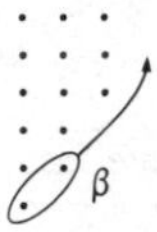

possible.  not possible.

β is possible when the number of nodes in the north-east slant line < smallest part, unless possibly the removal of the slant-line nodes changes the smallest part. β increases the number of parts by one, and so changes the parity of the partition.

40  $k<l$, α always possible.

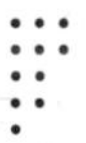 $k=l$, α sometimes possible.

 $k=l$, α impossible when slant line and smallest part overlap.

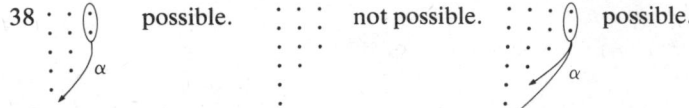

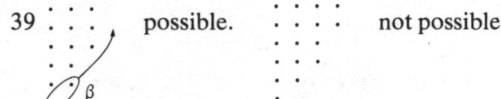

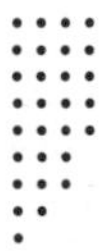 $k>l+1$, β always possible.

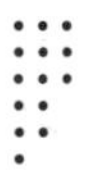 $k=l+1$, β sometimes possible.

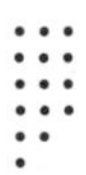 $k=l+1$, β impossible when slant line and smallest part overlap.

41, 42 Cases illustrated above.

43 If P is a partition of n into unequal parts with smallest part k and lowest north-east line of the Ferrers' graph containing l nodes, the transformation $T: P \to P'$ is well defined because $n \neq \frac{1}{2}l(3l \pm 1)$ for any l. If $k \leqslant l$ for P, then $k > l$ for P' and vice-versa. The mapping α reduces the number of parts by one and the mapping β increases the number of parts by one. If $T: P \to P'$ and $T: Q \to P'$, then either $k \leqslant l$ for both P and Q, or $k > l$ for both P and Q, so either $T = \alpha$ in both cases or $T = \beta$ in both cases, so P and Q have the same number of parts. If $k \leqslant l$ for both P and Q then (l for P') = (k for P) = (k for Q) so P and Q have the same smallest part, and, since only the smallest part is moved by T, $P = Q$. If $k > l$ for both P and Q, then (k for P') = (l for P) = (l for Q), and, since only these l nodes are moved by T, again $P = Q$. So T is one–one and matches each partition with an even number of unequal parts into a partition with an odd number of unequal parts and vice-versa. So $E(n) = O(n)$.

44 The cases where α and β are not possible, as in q 41, q 42 each have l parts. $(-1)^l = 1$ when l is even and -1 when l is odd so $E(n) - O(n) = 0$ unless $n = \frac{1}{2}l(3l \pm 1)$; then $E(n) - O(n) = (-1)^l$.

45 $1 - x - x^2 + x^5 + x^7 + \ldots$

46 The absolute values of the terms of the product arise from partitions into unequal parts. Those with an odd number of parts give a minus sign, those with an even number of parts give a plus sign.

47 $(1-x)(1-x^2)\ldots(1-x^n)\ldots$

$$= 1 + \sum_{n=1}^{\infty} (E(n) - O(n))x^n$$

$$= 1 + \sum_{l=1}^{\infty} (-1)^l x^{\frac{1}{2}l(3l-1)} + \sum_{l=1}^{\infty} (-1)^l x^{\frac{1}{2}l(3l+1)}$$

$$= 1 + \sum_{l=1}^{\infty} (-1)^l x^{\frac{1}{2}l(3l-1)} + \sum_{l=-1}^{-\infty} (-1)^{-l} x^{\frac{1}{2}(-l)(-3l+1)}$$

$$= 1 + \sum_{l=1}^{\infty} (-1)^l x^{\frac{1}{2}l(3l-1)} + \sum_{l=-1}^{-\infty} (-1)^l x^{\frac{1}{2}l(3l-1)}$$

$$= \sum_{l=-\infty}^{\infty} (-1)^l x^{\frac{1}{2}l(3l-1)}.$$

Historical note

The theory of partitions was first developed by L. Euler in the 1740s. Graphs of partitions were conceived by N. M. Ferrers and first appeared in print in a paper by J. J. Sylvester in 1853. The proof of Euler's theorem which we develop (q 7.36–q 7.47) is due to F. Franklin (in 1881). A comprehensive history of the subject is given in chapter 3 of volume 2 of Dickson (1950).

8

Quadratic forms

Unimodular transformations

We start by examining the transformations of a square lattice onto itself.

1 On some square lattice paper, choose four lattice points A, B, C and D so that $ABCD$ forms a parallelogram, Π, without lattice points inside the parallelogram or on its perimeter except at its vertices. Sketch the image of Π under the translation of the lattice which maps A to B. Call this translation τ. Sketch the image of Π under the translation of the lattice which maps A to D. Call this translation σ. Do $\tau(\Pi)$ and $\sigma(\Pi)$ have lattice points inside them or on their perimeters? Do $\tau^m(\Pi)$ or $\sigma^n(\Pi)$ have lattice points inside them or on their perimeters for any integers m or n? Is the same true for the parallelogram $\tau^m\sigma^n(\Pi)$?

2 With the notation of q 1, does every point of the plane lie within or on the perimeter of one or more of the parallelograms $\tau^m\sigma^n(\Pi)$? Must each lattice point be a vertex of four such parallelograms?

3 Using the conventional coordinate system with rectangular cartesian axes for the plane, the set of square lattice points may be labelled with the set of all ordered pairs of integers. If $ABCD$ is a parallelogram of lattice points as in q 1, we are free to choose A as the origin of our coordinate system. If $B = (a, b)$ and $D = (c, d)$, what are the coordinates of C if C is the vertex of the parallelogram opposite A? What is the image of the unit square $\{(0, 0), (1, 0), (1, 1), (0, 1)\}$ under the linear transformation

$$\alpha: (x, y) \to (x, y)\begin{pmatrix} a & b \\ c & d \end{pmatrix}?$$

What are the images of the lattice points of the form $(x, 0)$ under α? What are the images of the lattice points of the form $(0, y)$ under α? The lines $x = k$, $y = l$, parallel to the axes, with k and l integers, form a grid of unit squares. What is the image of this grid under the linear transformation α?

4 Does the linear transformation α of q 3, map the set of lattice points onto itself?

5 If $A = (0, 0)$, $B = (a, b)$, $C = (a + c, b + d)$ and $D = (c, d)$, find the area of the parallelogram $ABCD$.
If it happened that $ad - bc = 0$, what could you say about the relationship between the points A, B, C and D?

6 If under the linear transformation

$$\alpha: (x, y) \to (x, y)\begin{pmatrix} a & b \\ c & d \end{pmatrix}$$

there were two points (r, s) and (t, u) with the same image under α, prove that

$$\alpha: (r - t, s - u) \to (0, 0).$$

If $\alpha: (0, y) \to (0, 0)$ for $y \neq 0$, prove that $c = d = 0$.
If $\alpha: (x, y) \to (0, 0)$ for $x \neq 0$, prove that $ad - bc = 0$.
Deduce that if α is not one–one, then $ad - bc = 0$.

7 If the parallelogram $ABCD$ is not formed of collinear points, prove that the transformation α of q 3 is a one–one mapping of $\mathbf{Z}^2$ onto $\mathbf{Z}^2$.

8 With the conventions of q 3, must $ad - bc$ be an integer?
Find the point which is mapped to $(1, 0)$ under α.
Find the point which is mapped to $(0, 1)$ under α.
Must each of the numbers

$$\frac{a}{ad - bc}, \quad \frac{b}{ad - bc}, \quad \frac{c}{ad - bc}, \quad \frac{d}{ad - bc}$$

be an integer?
Prove that

$$\frac{a}{ad - bc} \cdot \frac{d}{ad - bc} - \frac{b}{ad - bc} \cdot \frac{c}{ad - bc}$$

is an integer and deduce that $ad - bc = \pm 1$.

⟦160⟧

9 If a, b, c and d are integers and $ad - bc = \pm 1$, does it follow that the transformation

$$(x, y) \to (x, y)\begin{pmatrix} a & b \\ c & d \end{pmatrix}$$

must map $\mathbf{Z}^2$ onto $\mathbf{Z}^2$?

10 What is the area of a parallelogram with lattice points as vertices, and with no other lattice points inside or on its sides?

11 Let M be the mid-point of the side BC of the triangle ABC and let A' be the image of A under a half-turn about M. What kind of quadrilateral is $ABA'C$? If A, B and C are lattice points of an infinite square lattice, must A' also be a point of the lattice? If there are no lattice points inside the triangle ABC, or on its sides (except at its vertices), can the same be said of $ABA'C$?

12 If A, B and C are lattice points of an infinite square lattice and the triangle ABC has no further lattice points inside or on its sides, what is its area?

Fig. 8.1

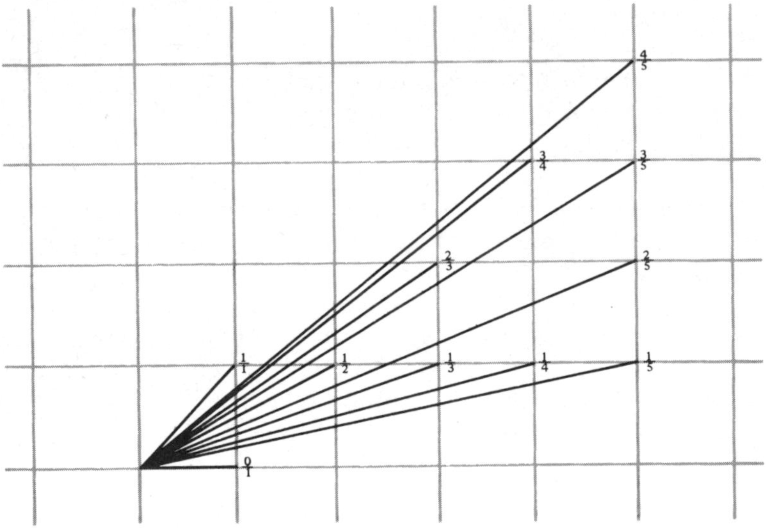

13 In the triangle with vertices (0, 0), (5, 0) and (5, 5) line segments are drawn from the origin to the lattice points inside or on the sides of the triangle. The slopes of the line segments which occur are

$0, \frac{1}{5}, \frac{1}{4}, \frac{1}{3}, \frac{2}{5}, \frac{1}{2}, \frac{3}{5}, \frac{2}{3}, \frac{3}{4}, \frac{4}{5}, 1,$

⟦161⟧

and these have been listed in strictly increasing order. This set of eleven numbers, in this order, is called the *Farey sequence* F_5. Write down the Farey sequence F_6, that is the set of rational numbers p/q between 0 and 1, where $0 \leqslant p \leqslant q \leqslant 6$, $\gcd(p, q) = 1$ and the terms are written down in ascending order.

14 If a/b, c/d, e/f, where a, b, c, d, e and f are positive integers, occur as consecutive terms in a Farey sequence, by considering the triangle with vertices $(0, 0)$, (b, a), (d, c), prove that $ad - bc = -1$.
Deduce that $cf - de = -1$ and that

$$\frac{c}{d} = \frac{a+e}{b+f}.$$

15 When $ps - qr = \pm 1$, and p, q, r and s are integers, the transformation

$$(x, y) \to (x, y)\begin{pmatrix} p & q \\ r & s \end{pmatrix}$$

is called a *unimodular transformation* and the matrix $\begin{pmatrix} p & q \\ r & s \end{pmatrix}$ a *unimodular matrix.*
What is the area of the image of a unit square under a unimodular transformation?

16 Is the transpose of a unimodular matrix unimodular?
Is the inverse of a unimodular matrix unimodular?

17 Prove that the set of unimodular matrices forms a group under matrix multiplication.
The corresponding group of unimodular transformations is known as the unimodular group. Sometimes it is convenient to think of this group as acting on the square lattice and sometimes on the whole plane.

18 If $\begin{pmatrix} p & q \\ r & s \end{pmatrix}$ is a unimodular matrix, and $X = px + ry$, $Y = qx + sy$, explain why $\gcd(x, y) | X, Y$. Prove that $x = \pm(sX - rY)$, $y = \pm(-qX + pY)$, and deduce that $\gcd(x, y) = \gcd(X, Y)$.

19 Each of the matrices

$$\begin{pmatrix} 1 & 0 \\ 0 & -1 \end{pmatrix}, \quad \begin{pmatrix} 0 & 1 \\ 1 & 0 \end{pmatrix}, \quad \begin{pmatrix} 0 & 1 \\ -1 & 0 \end{pmatrix}, \quad \begin{pmatrix} 1 & 0 \\ 1 & 1 \end{pmatrix}$$

is unimodular. Describe the effect of the corresponding unimodular transformations geometrically.

⟦162⟧

20 Identify a subgroup of index 2 in the unimodular group.
Does the subset of the unimodular group derived from the matrices of the form $\begin{pmatrix} 1 & 0 \\ a & 1 \end{pmatrix}$ form a subgroup?

Equivalent quadratic forms

21 If x and y are integers, what are the values less than 10 which x^2+y^2 can have?
On square lattice paper draw those circles $x^2+y^2=n$ which have lattice points on their circumference for $n<10$. What are the values of n for the circles you have drawn?

22 What values less than 10 can the expression

$$(x+y)^2+y^2=x^2+2xy+2y^2$$

adopt if x and y take only integral values?
Must all the values be positive?
Is 3 a possible value?

23 What values less than 10 can the quadratic form

$$(x+2y)^2+y^2=x^2+4xy+5y^2$$

adopt if x and y take only integral values?
For given integers m and n, is it possible to choose integers x and y so that

$$x+2y=m \quad \text{and} \quad y=n?$$

24 Using q 9, state conditions on p, q, r and s, under which the quadratic form

$$(px+ry)^2+(qx+sy)^2$$

adopts precisely the same set of integral values as the quadratic form x^2+y^2.

25 Because the transformation

$$(x, y) \rightarrow (x+y, y)$$

maps the integral lattice $\mathbf{Z}^2$ onto itself, the two quadratic forms x^2+y^2 and $(x+y)^2+y^2$ adopt precisely the same set of values and are said to be *equivalent* quadratic forms.
Prove that $(a+b, b)$ lies on $x^2+y^2=25$ if and only if (a, b) lies on $(x+y)^2+y^2=25$.
The transformation $(x, y) \rightarrow (x+y, y)=(x, y)\begin{pmatrix} 1 & 0 \\ 1 & 1 \end{pmatrix}$ thus maps the curve $(x+y)^2+y^2=25$ onto the circle $x^2+y^2=25$. Describe

the transformation $(x, y) \to (x, y)\begin{pmatrix} 1 & 0 \\ 1 & 1 \end{pmatrix}$ geometrically. Use this transformation to check the accuracy of the drawing of the ellipse $x^2+2xy+2y^2=25$ given in fig. 8.2.

Fig. 8.2

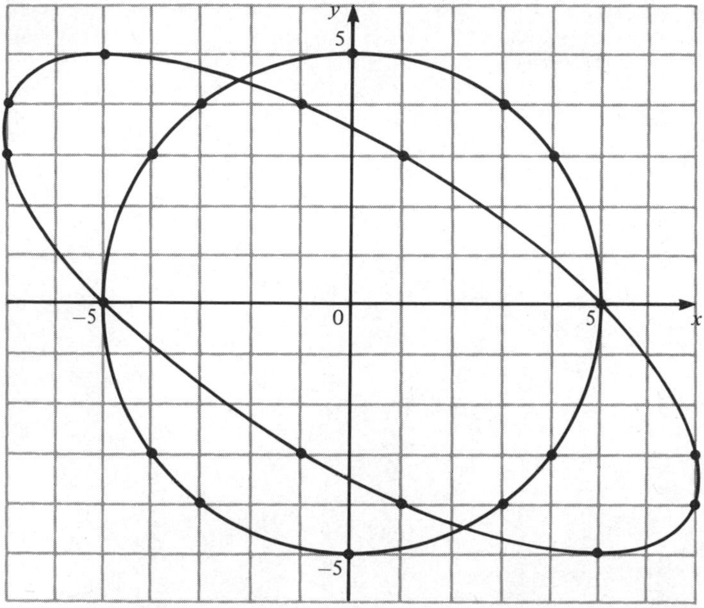

26 Into what curves are the three ellipses

$x^2+2xy+2y^2=9$

$x^2+2xy+2y^2=16$

and

$x^2+2xy+2y^2=25$

transformed by $(x, y) \to (x+y, y)$?

Sketch these three ellipses.

27 In table 8.1, the values adopted by the four quadratic forms

$x^2-2xy+3y^2$,

x^2+2y^2,

$x^2+2xy+3y^2$,

$x^2+4xy+6y^2$

are illustrated for integral values of x and y between -5 and $+5$. Prove that these four quadratic forms are equivalent.

⟦164⟧

Table 8.1

$x^2-2xy+3y^2$

$y=+5$	150	131	114	99	86	75	66	59	54	51	50
+4	113	96	81	68	57	48	41	36	33	32	33
+3	82	67	54	43	34	27	22	19	18	19	22
+2	57	44	33	24	17	12	9	8	9	12	17
+1	38	27	18	11	6	3	2	3	6	11	18
0	25	16	9	4	1	0	1	4	9	16	25
−1	18	11	6	3	2	3	6	11	18	27	38
−2	17	12	9	8	9	12	17	24	33	44	57
−3	22	19	18	19	22	27	34	43	54	67	82
−4	33	32	33	36	41	48	57	68	81	96	113
−5	50	51	54	59	66	75	86	99	114	131	150
$x=$	−5	−4	−3	−2	−1	0	1	2	3	4	5

x^2+2y^2

$y=+5$	75	66	59	54	51	50	51	54	59	66	75
+4	57	48	41	36	33	32	33	36	41	48	57
+3	43	34	27	22	19	18	19	22	27	34	43
+2	33	24	17	12	9	8	9	12	17	24	33
+1	27	18	11	6	3	2	3	6	11	18	27
0	25	16	9	4	1	0	1	4	9	16	25
−1	27	18	11	6	3	2	3	6	11	18	27
−2	33	24	17	12	9	8	9	12	17	24	33
−3	43	34	27	22	19	18	19	22	27	34	43
−4	57	48	41	36	33	32	33	36	41	48	57
−5	75	66	59	54	51	50	51	54	59	66	75
$x=$	−5	−4	−3	−2	−1	0	1	2	3	4	5

$x^2+2xy+3y^2$

$y=+5$	50	51	54	59	66	75	86	99	114	131	150
+4	33	32	33	36	41	48	57	68	81	96	113
+3	22	19	18	19	22	27	34	43	54	67	82
+2	17	12	9	8	9	12	17	24	33	44	57
+1	18	11	6	3	2	3	6	11	18	27	38
0	25	16	9	4	1	0	1	4	9	16	25
−1	38	27	18	11	6	3	2	3	6	11	18
−2	57	44	33	24	17	12	9	8	9	12	17
−3	82	67	54	43	34	27	22	19	18	19	22
−4	113	96	81	68	57	48	41	36	33	32	33
−5	150	131	114	99	86	75	66	59	54	51	50
$x=$	−5	−4	−3	−2	−1	0	1	2	3	4	5

Table 8.1 (*cont.*)

$x^2+4xy+6y^2$

$y=+5$	75	86	99	114	131	150	171	194	219	246	275
$+4$	41	48	57	68	81	96	113	132	153	176	201
$+3$	19	22	27	34	43	54	67	82	99	118	139
$+2$	9	8	9	12	17	24	33	44	57	72	89
$+1$	11	6	3	2	3	6	11	18	27	38	51
0	25	16	9	4	1	0	1	4	9	16	25
-1	51	38	27	18	11	6	3	2	3	6	11
-2	89	72	57	44	33	24	17	12	9	8	9
-3	139	118	99	82	67	54	43	34	27	22	19
-4	201	176	153	132	113	96	81	68	57	48	41
-5	275	246	219	194	171	150	131	114	99	86	75
$x=$	-5	-4	-3	-2	-1	0	1	2	3	4	5

28 If the shear

$$(x, y) \to (x, y)\begin{pmatrix}1 & 0\\ 1 & 1\end{pmatrix}$$

is expressed using column vectors as

$$\begin{pmatrix}x\\ y\end{pmatrix} \to \begin{pmatrix}p & q\\ r & s\end{pmatrix}\begin{pmatrix}x\\ y\end{pmatrix},$$

what are p, q, r and s?

29 $x^2+2y^2=(x \quad y)\begin{pmatrix}1 & 0\\ 0 & 2\end{pmatrix}\begin{pmatrix}x\\ y\end{pmatrix}.$

$$x^2+2xy+3y^2=(x \quad y)\begin{pmatrix}1 & 1\\ 1 & 3\end{pmatrix}\begin{pmatrix}x\\ y\end{pmatrix}.$$

Evaluate the matrix product

$$\begin{pmatrix}1 & 0\\ 1 & 1\end{pmatrix}\begin{pmatrix}1 & 0\\ 0 & 2\end{pmatrix}\begin{pmatrix}1 & 1\\ 0 & 1\end{pmatrix}$$

and deduce that (a, b) lies on $x^2+2xy+3y^2=k$ if and only if

$(a, b)\begin{pmatrix}1 & 0\\ 1 & 1\end{pmatrix}$ lies on $x^2+2y^2=k$.

30 As in q 24, x^2+2y^2 and $(px+ry)^2+2(qx+sy)^2$ are equivalent quadratic forms when p, q, r and s are integers and $ps-qr=\pm 1$. Find a matrix P such that

$$(px+ry)^2+2(qx+sy)^2=(x \quad y)P\begin{pmatrix}1 & 0\\ 0 & 2\end{pmatrix}P^{\mathrm{T}}\begin{pmatrix}x\\ y\end{pmatrix}$$

where P^{T} denotes the transpose of P.

⟦165⟧

31 $ax^2+bxy+cy^2=(x \quad y)\begin{pmatrix} a & \frac{1}{2}b \\ \frac{1}{2}b & c \end{pmatrix}\begin{pmatrix} x \\ y \end{pmatrix}$,

and

$a(px+ry)^2+b(px+ry)(qx+sy)+c(qx+sy)^2$

are equivalent quadratic forms when

$(x, y) \to (px+ry, qx+sy)$

is a unimodular transformation.

Find a matrix P such that

$$a(px+ry)^2+b(px+ry)(qx+sy)+c(qx+sy)^2$$
$$=(x \quad y)P\begin{pmatrix} a & \frac{1}{2}b \\ \frac{1}{2}b & c \end{pmatrix}P^{\mathrm{T}}\begin{pmatrix} x \\ y \end{pmatrix}.$$

Discriminant

We search for an easily recognised property which is shared by equivalent quadratic forms.

32 Express each of the quadratic forms

$x^2-2xy+3y^2$,

x^2+2y^2,

$x^2+2xy+3y^2$,

$x^2+4xy+6y^2$

in the form $(x \quad y)M\begin{pmatrix} x \\ y \end{pmatrix}$, for an appropriate choice of the symmetric matrix M.

Evaluate the determinant of each of these four matrices.

33 Use the working of q 17 to establish that the determinant of the product of two 2×2 matrices is equal to the product of the determinants of the matrices.

If P is a unimodular matrix and $A=\begin{pmatrix} a & b \\ c & d \end{pmatrix}$ prove that the determinant of PAP^{T} is $ad-bc$.

34 If A and B are symmetric matrices such that $(x \quad y)A\begin{pmatrix} x \\ y \end{pmatrix}$ and $(x \quad y)B\begin{pmatrix} x \\ y \end{pmatrix}$ are equivalent quadratic forms, prove that the determinants of A and B are equal.

35 If $ax^2+bxy+cy^2$ and $a'x^2+b'xy+c'y^2$ are equivalent quadratic forms, deduce a relationship between a, b, c, a', b' and c' from q 34.

⟦165⟧

36 Find the value of b^2-4ac for each of the three quadratic forms x^2+y^2, x^2 and x^2-y^2. Sketch the graphs of $x^2+y^2=1$, $x^2=1$ and $x^2-y^2=1$.

37 $$4a(ax^2+bxy+cy^2)=4a^2x^2+4abxy+4acy^2$$
$$=(2ax+by)^2+(4ac-b^2)y^2.$$

The number b^2-4ac is called the *discriminant* of the quadratic form $ax^2+bxy+cy^2$.
Use q 34 to verify that equivalent quadratic forms have the same discriminant.
Examine the possible values of the right hand side above when $b^2-4ac<0$, and prove that in this case $ax^2+bxy+cy^2$ must have the same sign as a.

38 When $b^2-4ac<0$, the quadratic form $ax^2+bxy+cy^2$ is said to be *definite*. If in addition $a>0$, it is said to be *positive definite*. If $a<0$, it is said to be *negative definite*.
If $ax^2+bxy+cy^2$ is positive definite, and $a'x^2+b'xy+c'y^2$ is an equivalent form, prove that $a'x^2+b'xy+c'y^2$ is also positive definite.

39 Give an example of a quadratic form which is not definite, and show that it either adopts both positive and negative values or else is zero for an infinity of pairs (x, y).

40 If a, b, c are integers, is it possible for b^2-4ac to equal -1 or -2? Consider the possibilities modulo 4, and try to generalise your result
(i) when b is even, (ii) when b is odd.
Give examples of quadratic forms for which $b^2-4ac=-3, -4, -7$ and -8.
Give an example of a quadratic form for which $b^2-4ac=-4k$.
Give an example of a quadratic form for which $b^2-4ac=-4k-3$.

41 If $b^2-4ac=-4$, determine the parity of b. List all the positive definite quadratic forms with this discriminant for which $|b|\leqslant 10$. Establish that each of these forms is equivalent to the form x^2+y^2.

42 If $b^2-4ac=-3$, determine the parity of b. List all the positive definite quadratic forms with this discriminant for which $|b|\leqslant 10$. Establish that each of these forms is equivalent to the form x^2+xy+y^2.

⟦166⟧

43 What is the discriminant of each of the forms $2x^2+3y^2$ and x^2+6y^2? Which of the numbers 1, 2, 3, 4, 5, 6 and 7 may be represented by which of these two forms? Are these two forms equivalent?

Proper representation

If $n=ax^2+bxy+cy^2$, then $k^2n=a(kx)^2+b(kx)(ky)+c(ky)^2$, so that if a quadratic form can represent a number, it can represent any square multiple of that number. Thus the values adopted by the form when $\gcd(x, y)=1$ imply all the remaining values which may be adopted.

44 Find the quadratic forms equivalent to x^2+y^2 obtained by substituting $(x, x+y)$, $(x, 2x+y)$, $(x, 3x+y)$, $(x, 4x+y)$ and $(2x+y, 3x+2y)$ respectively, for (x, y). List the coefficients of x^2 which occur in these forms. Compare this list with the possible values $\leqslant 20$ of the form x^2+y^2. What are the similarities and dissimilarities between the two lists?

45 If $(x, y)\rightarrow(px+ry, qx+sy)$ is a unimodular transformation, prove that the quadratic form obtained by substituting $(px+ry, qx+sy)$ for (x, y) in x^2+y^2 has a coefficient of x^2 which is itself one of the possible values of x^2+y^2.

46 If $(x, y)\rightarrow(px+ry, qx+sy)$ is a unimodular transformation, prove that $\gcd(p, q)=1$. If $\gcd(p, q)=1$, is it always possible to find r and s such that $(x, y)\rightarrow(px+ry, qx+sy)$ is a unimodular transformation? (See q 1.34.)

47 Determine whether 20, 26 and 29 can occur as the coefficient of x^2 in a quadratic form equivalent to x^2+y^2. If the coefficient can occur, give a substitution which transforms x^2+y^2 into such a form.

$1^2+0^2=1$,	$4^2+1^2=17$,
$1^2+1^2=2$,	$3^2+3^2=18=3^2(1^2+1^2)$,
$2^2+0^2=4=2^2(1^2+0^2)$,	$4^2+2^2=20=2^2(2^2+1^2)$,
$2^2+1^2=5$,	$4^2+3^2=25=5^2+0^2=5^2(1^2+0^2)$,
$2^2+2^2=8=2^2(1^2+1^2)$,	$5^2+1^2=26$,
$3^2+0^2=9=3^2(1^2+0^2)$,	$5^2+2^2=29$,
$3^2+1^2=10$,	$4^2+4^2=32=4^2(1^2+1^2)$,
$3^2+2^2=13$,	$5^2+3^2=34$,
$4^2+0^2=16=4^2(1^2+0^2)$,	$6^2+0^2=36=6^2(1^2+0^2)$.

[[168]]

48 If $\gcd(p, q) = 1$ and $n = ap^2 + bpq + cq^2$ then the integer n is said to be *properly represented* by the quadratic form $ax^2 + bxy + cy^2$. If p (or q) $= 0$, then n is properly represented by the form only when q (or p) $= \pm 1$.
List the numbers up to 20 which are properly represented by the form $x^2 + y^2$.

49 If n is properly represented by the form $ax^2 + bxy + cy^2$, prove that there is an equivalent form in which n is the coefficient of x^2.

50 If the quadratic form $ax^2 + bxy + cy^2$ is equivalent to the quadratic form $nx^2 + hxy + ly^2$, prove that n is properly represented by $ax^2 + bxy + cy^2$.

51 If $ax^2 + bxy + cy^2$ and $a'x^2 + b'xy + c'y^2$ are equivalent quadratic forms, so that there exists a unimodular matrix P such that

$$P\begin{pmatrix} a & \frac{1}{2}b \\ \frac{1}{2}b & c \end{pmatrix} P^{\mathrm{T}} = \begin{pmatrix} a' & \frac{1}{2}b' \\ \frac{1}{2}b' & c' \end{pmatrix},$$

and if moreover $n = ap^2 + bpq + cq^2$, with $\gcd(p, q) = 1$, find a representation for n by the form $a'x^2 + b'xy + c'y^2$, and use q 18 to prove that it is proper.

52 If the number 7 were to be representable by the quadratic form $x^2 + xy + 6y^2$ (and if it is representable, since it is prime, it is certainly properly representable) would there exist an equivalent quadratic form $7x^2 + hxy + ly^2$, for some h and l?
By considering the discriminant of these two forms modulo 7, establish that the representability of 7 by $x^2 + xy + 6y^2$ would lead to a contradiction.

Reduced forms

For positive definite quadratic forms we now determine a canonical form (a unique representative) for each class of equivalent forms.

53 Determine the discriminant of each of the forms $x^2 + xy + 4y^2$ and $2x^2 + xy + 2y^2$. Find the values which these forms adopt for $|x|, |y| < 3$, and conjecture whether or not these two forms are equivalent.
Let $f(x, y) = 2x^2 + xy + 2y^2$, and use the fact that $f(x, y) = 2(x + \frac{1}{4}y)^2 + \frac{15}{8}y^2$ to prove that $f(x, y) > 7$ when $|y| \geqslant 2$.
Prove further that $f(x, y) > 7$ when $|x| \geqslant 2$ and $y = \pm 1$.
Deduce that the least non-zero values which $2x^2 + xy + 2y^2$ properly represents are 2 and 3.
Prove that $x^2 + xy + 4y^2$ is not equivalent to $2x^2 + xy + 5y^2$.

⟦169⟧

54 Find the smallest non-zero numbers which are properly represented by the forms $2x^2+xy+5y^2$ and $3x^2+3xy+4y^2$, and deduce that they are not equivalent even though they have the same discriminant.

55 Show that the quadratic form $ax^2+bxy+cy^2$ takes the values a, c and $a+c-b$.
Show further that if $0\leqslant b\leqslant a\leqslant c$ and $0<a$, then when $|y|\geqslant 2$, $ax^2+bxy+cy^2\geqslant 3c>a+c$. Also, when $y=\pm 1$ and $|x|\geqslant 2$ show that $ax^2+bxy+cy^2\geqslant 2a+c>a+c$.
Deduce that the non-zero values $\leqslant a+c$ can only be properly represented when $(x, y)=(\pm 1, 0)$, $(0, \pm 1)$ or $(\pm 1, \pm 1)$, and that these values are a, c and $a+c-b$.

56 If the two forms $ax^2+bxy+cy^2$ and $a'x^2+b'xy+c'y^2$ are equivalent, $0\leqslant b\leqslant a\leqslant c$ and $0\leqslant b'\leqslant a'\leqslant c'$, prove that $a=a'$, $b=b'$ and $c=c'$.

57 When $0\leqslant b\leqslant a\leqslant c$, the positive definite quadratic form $ax^2+bxy+cy^2$ is said to be *reduced.*
Use the fact that $b^2\leqslant ac$ for a reduced form to prove that $ac\leqslant 1$ for a reduced quadratic form with discriminant -3. Find all the reduced quadratic forms with discriminant -3.

58 Find all the reduced quadratic forms with discriminant -4 and with discriminant -12.

59 Prove that $ac\leqslant\frac{1}{3}d$ for a reduced quadratic form with negative discriminant $-d$, and deduce that the number of reduced forms for any negative discriminant is finite.

60 By substituting $(x+y, y)$ for (x, y) a sufficient number of times, find a reduced form equivalent to $2x^2-11xy+18y^2$. Find a single substitution which establishes that $2x^2-11xy+18y^2$ is equivalent to a reduced form.

61 By substituting $(x-y, y)$ for (x, y) a sufficient number of times find a reduced form equivalent to $2x^2+13xy+24y^2$.
Find a single substitution which establishes that $2x^2+13xy+24y^2$ is equivalent to a reduced form.

62 For what value of the integer n does the substitution of $(x+ny, y)$ for (x, y) in the quadratic form

$$3x^2+50xy+211y^2$$

give the numerically least possible value for the coefficient of xy? Find a reduced form equivalent to the given form.

[[169]]

63 For what value of the integer n does the substitution of $(x+ny, y)$ for (x, y) in the quadratic form

$$3x^2+47xy+185y^2$$

give the numerically least possible value for the coefficient of xy? By further substituting (y, x) for (x, y) and $(-x, y)$ for (x, y) if necessary, find a reduced form equivalent to the given form.

64 Find a reduced form equivalent to the quadratic form $3x^2+46xy+177y^2$.

65 The substitution of $(x+ny, y)$ for (x, y) transforms the positive definite form $ax^2+bxy+cy^2$ to $a'x^2+b'xy+c'y^2$.
Prove that $a=a'$ and that it is possible to choose the integer n so that $|b'|\leq a$.

66 Explain how a positive definite quadratic form

$$ax^2+bxy+cy^2$$

can be shown to be equivalent to a reduced form by a sequence of (possibly repeated) substitutions of the types

$(x+ny, y)$ for (x, y), (y, x) for (x, y), and $(-x, y)$ for (x, y).

67 Prove that the number of equivalence classes of positive definite quadratic forms with a given discriminant is finite.

68 For which of the prime numbers $p=7$, 11, 13 and 17 is -2 a quadratic residue?
For those prime numbers p $(=7, 11, 13, 17)$ for which -2 is a quadratic residue, find values of h and l such that $h^2-pl=-2$, and then construct a quadratic form with discriminant -8 and p as the coefficient of x^2.
Determine all the reduced forms with discriminant -8, and deduce which of the primes 7, 11, 13 and 17 are representable by the form x^2+2y^2.

69 Determine the value of $\left(\frac{-2}{p}\right)$ for $p\equiv 1, 3, 5, 7 \pmod 8$. (See q 4.42.)
Deduce which primes are representable by the quadratic form x^2+2y^2.

70 Use the equation $(x^2+2y^2)(a^2+2b^2)=(xa+2yb)^2+2(xb-ya)^2$ to give a complete description of all the numbers representable by the form x^2+2y^2.

⟦171⟧

71 Determine all the reduced quadratic forms with discriminant -20. Construct a quadratic form for which the discriminant is -20 and the coefficient of x^2 is 30. Deduce that 30 is properly representable by one or other of the quadratic forms x^2+5y^2 or $2x^2+2xy+3y^2$.

72 If there exists exactly one reduced form with discriminant b^2-4ac, give a necessary and sufficient condition for the positive integer n to be properly represented by the form

$$ax^2+bxy+cy^2$$

in terms of the solubility of a quadratic congruence.

Automorphs of definite quadratic forms

73 For which unimodular transformations

$$(x, y) \to (px+ry, qx+sy)$$

is

$$(px+ry)^2+2(qx+sy)^2=x^2+2y^2?$$

Prove that these four unimodular transformations form a group, the group of *automorphs* of x^2+2y^2.

74 For which unimodular transformations

$$(x, y) \to (px+ry, qx+sy)$$

is

$$(px+ry)^2+(qx+sy)^2=x^2+y^2?$$

Prove that these eight unimodular transformations form a group, the group of automorphs of x^2+y^2.

75 For which unimodular transformations

$$(x, y) \to (px+ry, qx+sy)$$

is

$$(px+ry)^2+(px+ry)(qx+sy)+(qx+sy)^2=x^2+xy+y^2?$$

Prove that these twelve transformations form a group, the group of automorphs of x^2+xy+y^2.

76 What is the group of automorphs of the reduced quadratic form $ax^2+bxy+cy^2$,

(i) when $0=b<a<c$,
(ii) when $0<b\leqslant a<c$,
(iii) when $b=0$ and $a=c$,
(iv) when $0<b<a=c$,
(v) when $a=b=c$?

⟦173⟧

77 If A is a symmetric matrix and P and M are unimodular matrices, prove that $A = PAP^{\mathrm{T}}$ if and only if $(MPM^{-1})MAM^{\mathrm{T}}(MPM^{-1})^{\mathrm{T}} = MAM^{\mathrm{T}}$.
Deduce that the group of automorphs of any positive definite quadratic form is isomorphic to one of the groups of automorphs obtained in q 76.

Notes and answers

For concurrent reading see bibliography: Davenport (1968), Niven & Zuckerman (1972).

1 Because τ and σ are translations, $\tau^m\sigma^n(\Pi)$ is a parallelogram. Because τ and σ are translations of the lattice, the parallelogram $\tau^m\sigma^n(\Pi)$ has lattice points at its vertices. If there were lattice points inside $\tau^m\sigma^n(\Pi)$ or on its perimeter other than at its vertices, then their images under $\tau^{-m}\sigma^{-n}$ would be inside Π or on its perimeter.

2 The parallelograms $\tau^m(\Pi)$ cover an infinite strip unbounded in both directions. Because the direction of σ is not parallel to that of τ, the translations σ^n map this infinite strip onto parallel infinite strips to either side of the original strip without bound; so every point of the plane is covered by these strips. In particular, every lattice point is covered by these strips. But no lattice point lies within or on the perimeter of one of the parallelograms $\tau^m\sigma^n(\Pi)$ except at a vertex. Thus each lattice point is the image of either A, B, C or D under a translation $\tau^m\sigma^n$. If for example $\tau^m\sigma^n(A)=P$ then $\tau^{m-1}\sigma^n(B)=P$, $\tau^m\sigma^{n-1}(D)=P$, $\tau^{m-1}\sigma^{n-1}(C)=P$ if C is the vertex opposite A in the parallelogram $ABCD$, and then P is a vertex of the four parallelograms $\tau^m\sigma^n(\Pi)$, $\tau^{m-1}\sigma^n(\Pi)$, $\tau^m\sigma^{n-1}(\Pi)$ and $\tau^{m-1}\sigma^{n-1}(\Pi)$.

3 $C=(a+c, b+d)$.
Under α, the image of the unit square is the parallelogram $ABCD$. Under α, the images of the lattice points $(x, 0)$ are the lattice points on the line AB. Under α, the images of the lattice points $(0, y)$ are the lattice points on the line AD. The image of the grid of unit squares is the grid of unit parallelograms.

4 Yes, from q 2 and q 3.

5 By dropping perpendiculars from B, C and D to the x-axis and by calculating the areas of the right-angled triangles and trapezia formed, the area of the parallelogram is found to be the absolute value of $ad-bc$. This holds for any real numbers a, b, c and d. If the area of the parallelogram is zero, the four vertices are collinear.

The number $ad-bc$ is called the *determinant* of the matrix $\begin{pmatrix} a & b \\ c & d \end{pmatrix}$.

6 $(ar+cs, br+ds)=(at+cu, bt+du)$
$\Leftrightarrow (a(r-t)+c(s-u), b(r-t)+d(s-u))=(0, 0)$,

so

$\alpha: (r-t, s-u)\rightarrow(0, 0)$.
$\alpha: (0, y)\rightarrow(yc, yd)$ so if $y\neq 0$, $c=d=0$.

$(ax+cy, bx+dy)=(0, 0)$ with $x \neq 0$

$\Rightarrow d(ax+cy)-c(bx+dy)=0$

or $(ad-bc)x=0,$

and since $x \neq 0,$ $ad-bc=0.$

If α is not one–one, then there must be distinct points with the same image, and from the above working their difference is a point different from the origin which is mapped by α to the origin. Again we have shown whether this point lies on the y-axis or not, the conclusion $ad-bc=0$ follows.

7 α has already been proved to be onto in q 4. If α were not one–one, then $ad-bc=0$ from q 6 and the four points A, B, C, D would be collinear.

8 If a, b, c, and d are integers, then so is $ad-bc$.

$$\alpha:\left(\frac{d}{ad-bc}, \frac{-b}{ad-bc}\right) \to (1, 0).$$

$$\alpha:\left(\frac{-c}{ad-bc}, \frac{a}{ad-bc}\right) \to (0, 1).$$

But α is a one–one mapping of $\mathbf{Z}^2$ onto itself, so the four coordinates here are integers. Hence their products and sums are integers and so

$$\frac{a}{ad-bc}\cdot\frac{d}{ad-bc}-\frac{b}{ad-bc}\cdot\frac{c}{ad-bc}=\frac{ad-bc}{(ad-bc)^2}=\frac{1}{ad-bc}$$

is an integer. But the only integers which are the reciprocals of integers are ± 1, so $ad-bc=\pm 1$.

9 If a, b, c, d are integers, the transformation maps $\mathbf{Z}^2$ into $\mathbf{Z}^2$. If $ad-bc=\pm 1$ then the matrix inverse to $\begin{pmatrix} a & b \\ c & d \end{pmatrix}$, namely

$$\begin{pmatrix} \dfrac{d}{ad-bc} & \dfrac{-b}{ad-bc} \\ \dfrac{-c}{ad-bc} & \dfrac{a}{ad-bc} \end{pmatrix},$$

also has integer entries, so each lattice point has a lattice point as pre-image, and the transformation of $\mathbf{Z}^2$ is onto.

10 The argument of q 1–8 establishes that the area of such a parallelogram is 1.

11 A half-turn maps each line to a parallel line, and the half-turn about M interchanges B and C, so $ABA'C$ is a parallelogram. If the lattice is labelled with ordered pairs of integers in the conventional way then the coordinates of M are either integers or integers plus one half. The half-turn about (r, s) is $(x, y) \to (-x+2r, -y+2s)$ which certainly maps $\mathbf{Z}^2$

onto itself if $M=(r,s)$; so A' is a lattice point. Since the half-turn maps the lattice onto itself, the only lattice points in the triangle $A'BC$ are the images of lattice points in ABC under the half-turn.

12 Half a unit. The construction of 11 doubles the triangle to form a parallelogram of unit area.

13 $0, \frac{1}{6}, \frac{1}{5}, \frac{1}{4}, \frac{1}{3}, \frac{2}{5}, \frac{1}{2}, \frac{3}{5}, \frac{2}{3}, \frac{3}{4}, \frac{4}{5}, \frac{5}{6}, 1.$

14 If a/b is a term of a Farey sequence, then $\gcd(a,b)=1$ and so there are no lattice points on the line segment joining $(0,0)$ to (b,a). Likewise there can be no lattice point on the line segment joining $(0,0)$ to (d,c). Since a/b and c/d are adjacent terms of a Farey sequence there is no lattice point on the line segment joining (b,a) to (d,c) or within the triangle with vertices $(0,0)$, (b,a), (d,c). Therefore this triangle has area half a unit as in q 12 and by q 5 $ad-bc=\pm 1$.
But $a/b<c/d$ so $ad<bc$ and $ad-bc=-1$.
Similarly $cf-de=-1$.
So $ad-bc=cf-de$ and $d(a+e)=c(b+f)$.

15 The unimodular transformations are the automorphisms of a square lattice which fix one of the lattice points. The image of the unit square $(0,0)$, $(1,0)$, $(1,1)$, $(0,1)$ under such a transformation is $(0,0)$, (p,q), $(p+r,q+s)$, (r,s), which form a parallelogram with unit area.
The authorities are almost equally divided between those who use our definition of unimodular and those who only apply the name to a matrix with determinant $+1$.

16 The transpose of $\begin{pmatrix} p & q \\ r & s \end{pmatrix}$ is $\begin{pmatrix} p & r \\ q & s \end{pmatrix}$ and both have the same determinant.
So either both or neither are unimodular.
The inverse of $\begin{pmatrix} p & q \\ r & s \end{pmatrix}$ is $\begin{pmatrix} \pm s & \mp q \\ \mp r & \pm p \end{pmatrix}$ when $ps-qr=\pm 1$, and this again is unimodular.

17 Obviously the unit matrix is unimodular so, with q 16, we have only to prove closure.
Let $ps-qr=\pm 1$ and $ad-bc=\pm 1$, then

$$\begin{pmatrix} p & q \\ r & s \end{pmatrix}\begin{pmatrix} a & b \\ c & d \end{pmatrix}=\begin{pmatrix} pa+qc & pb+qd \\ ra+sc & rb+sd \end{pmatrix}.$$

The determinant of this last matrix is

$$\begin{aligned}&(pa+qc)(rb+sd)-(ra+sc)(pb+qd)\\ &\quad = ps\cdot ad+qr\cdot bc+pr\cdot ab+qs\cdot cd-pr\cdot ab-ps\cdot bc-qr\cdot ad-qs\cdot cd\\ &\quad = ps\cdot ad+qr\cdot bc-ps\cdot bc-qr\cdot ad\\ &\quad =(ps-qr)(ad-bc)\\ &\quad =\pm 1.\end{aligned}$$

18 gcd $(x, y) = $ gcd (X, Y) because each divides the other. Geometrically, gcd (x, y) is the number of lattice points on the line segment joining $(0, 0)$ to (x, y), excluding the point $(0, 0)$. Under a unimodular transformation, a set of collinear lattice points is mapped one–one onto a set of collinear lattice points.

19 $(x, y) \to (x, y)\begin{pmatrix} 1 & 0 \\ 0 & -1 \end{pmatrix}$ is a reflection in the x-axis.

$(x, y) \to (x, y)\begin{pmatrix} 0 & 1 \\ 1 & 0 \end{pmatrix}$ is a reflection in $y = x$.

$(x, y) \to (x, y)\begin{pmatrix} 0 & 1 \\ -1 & 0 \end{pmatrix}$ is an anti-clockwise quarter-turn about the origin.

$(x, y) \to (x, y)\begin{pmatrix} 1 & 0 \\ 1 & 1 \end{pmatrix}$ is a shear on the x-axis.

20 The matrices with determinant +1 give a subgroup of index 2.

$$\begin{pmatrix} 1 & 0 \\ a & 1 \end{pmatrix}\begin{pmatrix} 1 & 0 \\ b & 1 \end{pmatrix} = \begin{pmatrix} 1 & 0 \\ a+b & 1 \end{pmatrix},$$

so these matrices form a group isomorphic to $(\mathbf{Z}, +)$. The corresponding transformations are shears on the x-axis.

21 $1^2+0^2=1$, $\quad 3^2+0^2=9$,
$1^2+1^2=2$, $\quad 3^2+1^2=10$,
$2^2+0^2=4$, $\quad 3^2+2^2=13$,
$2^2+1^2=5$, $\quad 3^2+3^2=18$.
$2^2+2^2=8$,

See fig. 8.3; $n = 1, 2, 4, 5, 8, 9$.

Fig. 8.3

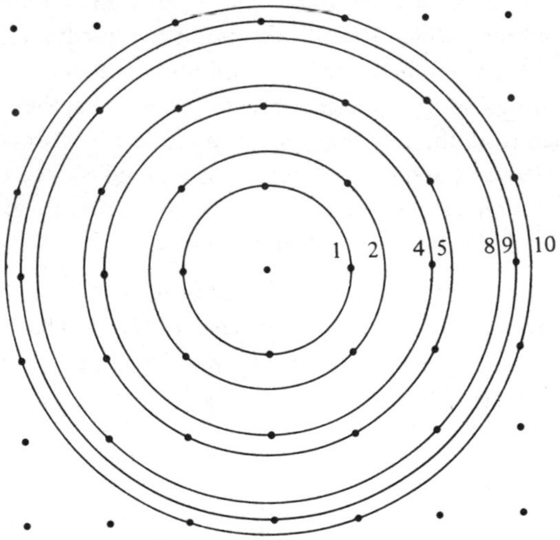

22 Certainly $(x+y)^2$ and y^2 can only be positive or zero. Since $x+y$ and y are both integers only the values adopted in q 21 may occur. The number 3 is not the sum of two squares.

23 If $x=m-2n$ and $y=n$, we have $x+2y=m$ and $y=n$. Thus $(x+2y)^2+y^2$ has the same values as m^2+n^2 or x^2+y^2.
For given a, b, and c, a function of x and y such that $f(x, y)=ax^2+bxy+cy^2$ is called a *binary quadratic form.*

24 If $ps-qr=\pm 1$ and p, q, r and s are integers there is a lattice point (x, y) which is mapped onto the lattice point $(px+ry, qx+sy)$ by the unimodular transformation

$$(x, y) \rightarrow (x, y)\begin{pmatrix} p & q \\ r & s \end{pmatrix}$$

Thus $px+ry$ and $qx+sy$ can simultaneously take any pair of integral values by means of a suitable choice of x and y.

25 $(a+b, b)$ lies on $x^2+y^2=25 \Leftrightarrow (a+b)^2+b^2=25$

$$\Leftrightarrow (a, b) \text{ lies on } (x+y)^2+y^2=25$$

$(x, y) \rightarrow (x+y, y)$ is a shear fixing each point on the x-axis and moving each point parallel to the x-axis through a distance equal to its distance from the x-axis.
While, for the purpose of number theory, we are only concerned with integral values of the variables x and y, and thus only with a small number of points on the circle $x^2+y^2=25$, yet the drawing of circles and ellipses in the conventional way can give visual models of some quadratic forms which are not misleading if we remember that our concern is principally with the lattice points.
Two quadratic forms are defined to be *equivalent* when a unimodular transformation can be found which transforms one of the quadratic forms into the other. One consequence of this definition is that equivalent quadratic forms adopt the same integral values. Another consequence, which depends on the fact that the unimodular transformations form a group is that the equivalence thus defined is an equivalence relation in the usual sense, and so this relation of equivalence partitions the set of all binary quadratic forms into disjoint classes.

26 $x^2+2xy+2y^2=(x+y)^2+y^2$, so (a, b) lies on $x^2+2xy+2y^2=k$ when $(a+b, b)$ lies on $x^2+y^2=k$. Thus the three ellipses are transformed into the three circles $x^2+y^2=9$, $x^2+y^2=16$ and $x^2+y^2=25$ respectively. Make the sketches by shearing the circles.

27 $x^2-2xy+3y^2=(x-y)^2+2y^2$,
$x^2+2xy+3y^2=(x+y)^2+2y^2$,
$x^2+4xy+6y^2=(x+2y)^2+2y^2$.

The three transformations

$(x, y) \to (x-y, y)$,

$(x, y) \to (x+y, y)$,

$(x, y) \to (x+2y, y)$

are all unimodular and thus establish that each of the three forms given above is equivalent to x^2+2y^2.

28 $\begin{pmatrix}1 & 1\\0 & 1\end{pmatrix}\begin{pmatrix}x\\y\end{pmatrix}=\begin{pmatrix}x+y\\y\end{pmatrix}$.

29 $\begin{pmatrix}1 & 0\\1 & 1\end{pmatrix}\begin{pmatrix}1 & 0\\0 & 2\end{pmatrix}\begin{pmatrix}1 & 1\\0 & 1\end{pmatrix}=\begin{pmatrix}1 & 0\\1 & 2\end{pmatrix}\begin{pmatrix}1 & 1\\0 & 1\end{pmatrix}=\begin{pmatrix}1 & 1\\1 & 3\end{pmatrix}$.

(a, b) lies on $x^2+2xy+3y^2=k$

$$\Leftrightarrow (a \quad b)\begin{pmatrix}1 & 1\\1 & 3\end{pmatrix}\begin{pmatrix}a\\b\end{pmatrix}=k$$

or

$$(a \quad b)\begin{pmatrix}1 & 0\\1 & 1\end{pmatrix}\begin{pmatrix}1 & 0\\0 & 2\end{pmatrix}\begin{pmatrix}1 & 1\\0 & 1\end{pmatrix}\begin{pmatrix}a\\b\end{pmatrix}=k$$

or

$$(a, b)\begin{pmatrix}1 & 0\\1 & 1\end{pmatrix} \text{ lies on } (x \quad y)\begin{pmatrix}1 & 0\\0 & 2\end{pmatrix}\begin{pmatrix}x\\y\end{pmatrix}=k.$$

30 $(px+ry)^2+2(qx+sy)^2=(x \quad y)\begin{pmatrix}p & q\\r & s\end{pmatrix}\begin{pmatrix}1 & 0\\0 & 2\end{pmatrix}\begin{pmatrix}p & r\\q & s\end{pmatrix}\begin{pmatrix}x\\y\end{pmatrix}$.

31 $P=\begin{pmatrix}p & q\\r & s\end{pmatrix}$.

This result expresses in a general algebraic way how equivalent quadratic forms are linked by unimodular transformations.

32 $x^2-2xy+3y^2=(x \quad y)\begin{pmatrix}1 & -1\\-1 & 3\end{pmatrix}\begin{pmatrix}x\\y\end{pmatrix}$, determinant $=2$.

$x^2+2y^2=(x \quad y)\begin{pmatrix}1 & 0\\0 & 2\end{pmatrix}\begin{pmatrix}x\\y\end{pmatrix}$, determinant $=2$.

$x^2+2xy+3y^2=(x \quad y)\begin{pmatrix}1 & 1\\1 & 3\end{pmatrix}\begin{pmatrix}x\\y\end{pmatrix}$, determinant $=2$.

$x^2+4xy+6y^2=(x \quad y)\begin{pmatrix}1 & 2\\2 & 6\end{pmatrix}\begin{pmatrix}x\\y\end{pmatrix}$, determinant $=2$.

33 In n 17 the condition that the matrices are unimodular is imposed only after establishing the result we want here. Hence det $PAP^{\mathrm{T}}=$ (det P)(det A)(det P^{T}). Now det $P=$ det P^{T} for all P, and if P is unimodular, det $P=\pm 1$. Thus (det P)(det P^{T}) $=1$ and det $PAP^{\mathrm{T}}=$ det A.

34 If these forms are equivalent, then for some unimodular P, $B = PAP^{\mathrm{T}}$ and by q 33, $\det B = \det A$.

35 $ax^2 + bxy + cy^2 = (x \quad y)\begin{pmatrix} a & \frac{1}{2}b \\ \frac{1}{2}b & c \end{pmatrix}\begin{pmatrix} x \\ y \end{pmatrix}$. Determinant of the 2×2 symmetric matrix is $ac - \frac{1}{4}b^2$. Thus $ac - \frac{1}{4}b^2 = a'c' - \frac{1}{4}b'^2$, or $b^2 - 4ac = b'^2 - 4a'c'$.

36 $x^2 + y^2$, $b^2 - 4ac = -4$, $x^2 + y^2 = 1$ is a circle.
x^2, $b^2 - 4ac = 0$, $x^2 = 1$ is a pair of parallel lines.
$x^2 - y^2$, $b^2 - 4ac = 4$, $x^2 - y^2 = 1$ is a rectangular hyperbola.

The point of this exercise is to illustrate the difference between quadratic forms as the sign of $b^2 - 4ac$ changes.

37 The discriminant $= -4 \times$ the determinant of the corresponding matrix. If $b^2 - 4ac < 0$, $4ac - b^2 > 0$ so $(4ac - b^2)y^2$ is positive or zero. Thus the right hand side is positive or zero, so both $4a$ and $ax^2 + bxy + cy^2$ must have the same sign.

38 A positive definite quadratic form always takes positive or zero values, and corresponds to a family of concentric ellipses.
A negative definite quadratic form always takes negative or zero values. Equivalent forms have the same discriminant, so a form equivalent to a definite quadratic form is definite. Moreover, equivalent forms adopt the same integral values, so either both are positive or both negative. Thus the equivalence of quadratic forms as defined previously is also an equivalence relation on the set of positive definite quadratic forms.

39 $x^2 - y^2$ is positive when $y = 0$ and negative when $x = 0$.
x^2 is zero for all $(0, y)$.

40 If b is even $b^2 \equiv 0 \pmod 4$ so $b^2 - 4ac \equiv 0 \pmod 4$.
If b is odd $b^2 \equiv 1 \pmod 4$ so $b^2 - 4ac \equiv 1 \pmod 4$.
Neither -1 nor $-2 \equiv 0$ or $1 \pmod 4$. Since $b^2 - 4ac \equiv 0$ or $1 \pmod 4$ for all a, b and c, it follows that the discriminant cannot equal $-1, -2, -5, -6, -9, -10$, etc.

$b^2 - 4ac = -3$	$x^2 + xy + y^2$
$b^2 - 4ac = -4$	$x^2 + y^2$
$b^2 - 4ac = -7$	$x^2 + xy + 2y^2$
$b^2 - 4ac = -8$	$x^2 + 2y^2$
$b^2 - 4ac = -4k$	$x^2 + ky^2$
$b^2 - 4ac = -4k - 3$	$x^2 + xy + (k+1)y^2$.

41 If $b^2 - 4ac = -4$, $b^2 = 4(ac - 1)$ so b is even.
$b = 0 \Rightarrow ac = 1 \Rightarrow a = 1$, $\quad x^2 + y^2$

$b=\pm 2 \Rightarrow ac=2$ $\quad x^2 \pm 2xy+2y^2=(x\pm y)^2+y^2,$

$\Rightarrow a=1$ or 2 $\quad 2x^2 \pm 2xy+y^2=x^2+(x\pm y)^2.$

$b=\pm 4 \Rightarrow ac=5$ $\quad x^2 \pm 4xy+5y^2=(x\pm 2y)^2+y^2,$

$\Rightarrow a=1$ or 5 $\quad 5x^2 \pm 4xy+y^2=x^2+(x\pm 2y)^2.$

$b=\pm 6 \Rightarrow ac=10$ $\quad x^2 \pm 6xy+10y^2=(x\pm 3y)^2+y^2,$

$\Rightarrow a=1, 2, 5, 10$ $\quad 2x^2 \pm 6xy+5y^2=(x\pm y)^2+(x\pm 2y)^2,$

$5x^2 \pm 6xy+2y^2=(x\pm y)^2+(2x\pm y)^2,$

$10x^2 \pm 6xy+y^2=x^2+(3x\pm y)^2.$

$b=\pm 8 \Rightarrow ac=17$ $\quad x^2 \pm 8xy+17y^2=(x\pm 4y)^2+y^2,$

$\Rightarrow a=1, 17$ $\quad 17x^2 \pm 8xy+y^2=x^2+(4x\pm y)^2.$

$b=\pm 10 \Rightarrow ac=26$ $\quad x^2 \pm 10xy+26y^2=(x\pm 5y)^2+y^2,$

$\Rightarrow a=1, 2, 13, 26$ $\quad 2x^2 \pm 10xy+13y^2=(x\pm 2y)^2+(x\pm 3y)^2,$

$13x^2 \pm 10xy+2y^2=(2x\pm y)^2+(3x\pm y)^2,$

$26x^2 \pm 10xy+y^2=x^2+(5x\pm y)^2.$

42 $b^2=4ac-3$ so b is odd.

$b=\pm 1 \Rightarrow ac=1$ $\quad x^2 \pm xy+y^2=x^2+x(\pm y)+(\pm y)^2.$

$\Rightarrow a=1$

$b=\pm 3 \Rightarrow ac=3$ $\quad x^2 \pm 3xy+3y^2=(x\pm y)^2+(x\pm y)(\pm y)+(\pm y)^2,$

$\Rightarrow a=1$ or 3 $\quad 3x^2 \pm 3xy+y^2=(\pm x)^2+(\pm x)(x\pm y)+(x\pm y)^2.$

$b=\pm 5 \Rightarrow ac=7$ $\quad x^2 \pm 5xy+7y^2$

$\Rightarrow a=1$ or 7 $\quad =(x\pm 2y)^2+(x\pm 2y)(\pm y)+(\pm y)^2,$

$7x^2 \pm 5xy+y^2$

$=(\pm x)^2+(\pm x)(2x\pm y)+(2x\pm y)^2.$

$b=\pm 7 \Rightarrow ac=13$ $\quad x^2 \pm 7xy+13y^2$

$\Rightarrow a=1$ or 13 $\quad =(x\pm 3y)^2+(x\pm 3y)(\pm y)+(\pm y)^2,$

$13x^2 \pm 7xy+y^2$

$=(\pm x)^2+(\pm x)(3x\pm y)+(3x\pm y)^2.$

$b=\pm 9 \Rightarrow ac=21$ $\quad x^2 \pm 9xy+21y^2$

$\Rightarrow a=1, 3, 7$ or 21 $\quad =(x\pm 4y)^2+(x\pm 4y)(\pm y)+(\pm y)^2,$

$3x^2 \pm 9xy+7y^2$

$=(x\pm 2y)^2+(x\pm 2y)(x\pm y)+(x\pm y)^2,$

$7x^2 \pm 9xy+3y^2$

$=(x\pm y)^2+(x\pm y)(2x\pm y)+(2x\pm y)^2,$

$21x^2 \pm 9xy+y^2$

$=(\pm x)^2+(\pm x)(4x\pm y)+(4x\pm y)^2.$

43 $2x^2+3y^2$ can take the values 2, 3 and 5, but not 1, 4, 6 or 7.
x^2+6y^2 can take the values 1, 4, 6 and 7, but not 2, 3 or 5.
Although these two forms both have discriminant −24, they cannot be equivalent as they do not adopt the same set of integral values.
Thus having the same discriminant is a necessary but not a sufficient condition for the equivalence of two quadratic forms.

44 Coefficients of $x^2 = 2, 5, 10, 17, 13$.
Possible values of $x^2+y^2 = 1, 2, 4, 5, 8, 9, 10, 13, 16, 17, 18, 20$.
The list of coefficients is a subset of the list of possible values.

45 $(px+ry)^2+(qx+sy)^2=(p^2+q^2)x^2+2(pr+qs)xy+(r^2+s^2)y^2$.

46 If the transformation is unimodular, then $ps-qr=\pm 1$ and so $\gcd(p, q) = 1$. If $\gcd(p, q)=1$, then integers s and r may be found such that $ps-qr=1$ from q 1.34, and then $(x, y)\to(px+ry, qx+sy)$ is unimodular.

47 $(5x+4y)^2+(x+y)^2=26x^2+42xy+17y^2$,

$(5x+2y)^2+(2x+y)^2=29x^2+24xy+5y^2$.

48 1, 2, 5, 10, 13, 17.

49 If n is properly represented by $ax^2+bxy+cy^2$, then $n=ap^2+bpq+cq^2$ for $\gcd(p, q)=1$. Now from q 46, there is a unimodular transformation $(x, y)\to(px+ry, qx+sy)$, and so $ax^2+bxy+cy^2$ is equivalent to

$$\begin{aligned} &a(px+ry)^2+b(px+ry)(qx+sy)+c(qx+sy)^2\\ &\quad=(ap^2+bpq+cq^2)x^2+\ldots\\ &\quad=nx^2+\ldots \end{aligned}$$

50 If $ax^2+bxy+cy^2$ is equivalent to $nx^2+hxy+ly^2$ then there exists a unimodular transformation

$$(x, y)\to(px+ry, qx+sy)$$

such that

$$a(px+ry)^2+b(px+ry)(qx+sy)+c(qx+sy)^2=nx^2+hxy+ly^2$$

so $n=ap^2+bpq+cq^2$.
Since the transformation is unimodular, $\gcd(p, q)=1$, and n is properly represented.

51 Since

$$P\begin{pmatrix} a & \frac{1}{2}b \\ \frac{1}{2}b & a \end{pmatrix}P^{\mathrm{T}}=\begin{pmatrix} a' & \frac{1}{2}b' \\ \frac{1}{2}b' & c' \end{pmatrix},$$

$$\begin{pmatrix} a & \frac{1}{2}b \\ \frac{1}{2}b & c \end{pmatrix}=P^{-1}\begin{pmatrix} a' & \frac{1}{2}b' \\ \frac{1}{2}b' & c' \end{pmatrix}P^{-1\mathrm{T}},$$

so

$$n=(p\quad q)\begin{pmatrix} a & \frac{1}{2}b \\ \frac{1}{2}b & c \end{pmatrix}\begin{pmatrix} p \\ q \end{pmatrix}=(p\quad q)P^{-1}\begin{pmatrix} a' & \frac{1}{2}b' \\ \frac{1}{2}b' & c' \end{pmatrix}P^{-1\mathrm{T}}\begin{pmatrix} p \\ q \end{pmatrix}.$$

Now if $(p, q)P^{-1} = (X, Y)$, $n = a'X^2 + b'XY + c'Y^2$, which is a proper representation by q 18.

52 From q 49 an equivalent form $7x^2 + hxy + ly^2$ would have to exist. Equivalent forms have equal discriminant, so $h^2 - 28l = 1 - 24$ or $h^2 = 5 + 7(4l - 4)$ so $h^2 \equiv 5 \pmod 7$. But 5 is not a quadratic residue modulo 7.

53 The discriminant is -15 in each case.
Let $g(x, y) = x^2 + xy + 4y^2$ and $f(x, y) = 2x^2 + xy + 2y^2$.
For $|x|$, $|y| < 3$, $g(x, y) = 0, 1, 4, 6, 10, 15, 16, 19, 24$ and $f(x, y) = 0, 2, 3, 5, 8, 12, 20$.
When $|y| \geqslant 2$, $\frac{15}{8}y^2 \geqslant \frac{15}{2}$, so $f(x, y) > 7$.
When $y = \pm 1$, $f(x, y) = 2(x \pm \frac{1}{4})^2 + \frac{15}{8}$ and so if $|x| \geqslant 2$, $f(x, y) \geqslant 2(\frac{7}{4})^2 + \frac{15}{8} = 8$.
When $y = 0$, for proper representation $x = \pm 1$. Thus for $f(x, y) < 8$, both $|x|$ and $|y|$ are less than 2 and these cases were enumerated at the start.
$g(x, y) = (x + \frac{1}{2}y)^2 + \frac{15}{4}y^2$, so $g(x, y) \geqslant 15$ when $|y| \geqslant 2$.
When $y = \pm 1$, $g(x, y) = (x \pm \frac{1}{2})^2 + \frac{15}{4}$, so if in addition $|x| \geqslant 2$, $g(x, y) \geqslant (\frac{3}{2})^2 + \frac{15}{4} = 6$.
When $y = 0$, $x = \pm 1$ for proper representation thus, for properly represented values of $g(x, y) < 6$, both $|x|$ and $|y|$ are less than 2, and these cases were enumerated at the start. This proves that the smallest non-zero values properly represented by f and g are different, so the forms cannot be equivalent.

54 Let $f(x, y) = 2x^2 + xy + 5y^2 = 2(x + \frac{1}{4}y)^2 + \frac{39}{8}y^2$.
When $|y| \geqslant 2$, $f(x, y) \geqslant \frac{39}{2}$.
When $|x| \geqslant 2$, $f(x, \pm 1) = 2(x \pm \frac{1}{4})^2 + \frac{39}{8} \geqslant 2(\frac{7}{4})^2 + \frac{39}{8} = 11$.
When $y = 0$, $x = \pm 1$ for proper representation thus, for properly represented $f(x, y) < 11$, both $|x|$ and $|y|$ are less than 2.

$f(\pm 1, 0) = 2$, $f(0, \pm 1) = 5$, $f(\pm 1, \pm 1) = 8$, $f(\pm 1, \mp 1) = 6$.

Let $g(x, y) = 3x^2 + 3xy + 4y^2 = 3(x + \frac{1}{2}y)^2 + \frac{13}{4}y^2$. So when $|y| \geqslant 2$, $g(x, y) \geqslant 13$.
When $|x| \geqslant 2$, $g(x, \pm 1) = 3(x \pm \frac{1}{2})^2 + \frac{13}{4} \geqslant 3(\frac{3}{2})^2 + \frac{13}{4} = 10$.
When $y = 0$, $x = \pm 1$ for proper representation thus, for properly represented $g(x, y) < 10$, both $|x|$ and $|y|$ are less than 2.

$g(\pm 1, 0) = 3$, $g(0, \pm 1) = 4$, $g(\pm 1, \pm 1) = 10$, $g(\pm 1, \mp 1) = 4$.

This proves that the smallest non-zero values adopted by f and g are different, so the forms cannot be equivalent.

55 Let $f(x, y) = ax^2 + bxy + cy^2$. Then $f(1, 0) = a$, $f(0, 1) = c$ and $f(1, -1) = a + c - b$.
If $0 \leqslant b \leqslant a \leqslant c$, $0 < a$ and $|y| \geqslant 2$,

$$f(x, y) = a\left(x + \frac{b}{2a}y\right)^2 + \left(c - \frac{b^2}{4a}\right)y^2 \geqslant \left(c - \frac{b^2}{4a}\right)4.$$

Now $b^2/a \leqslant c$ so $f(x, y) \geqslant 3c > a + c$.

When $|x| \geqslant 2$, $f(x, \pm 1) = a\left(x \pm \dfrac{b}{2a}\right)^2 + c - \dfrac{b^2}{4a} \geqslant a(\tfrac{3}{2})^2 + c - \tfrac{1}{4}b \geqslant 2a + c$.

Since $f(x, 0)$ is not a proper representation if $|x| \geqslant 2$, for numbers less than $a + c$ properly represented by f, both $|x|$ and $|y|$ are less than 2.

$f(\pm 1, 0) = a$, $f(0, \pm 1) = c$, $f(\pm 1, \pm 1) = a + b + c$, $f(\pm 1, \mp 1) = a - b + c$.

56 If two forms are equivalent, they properly represent the same values, from q 51, and hence the same smallest values. Thus, from q 55, $a = a'$, $c = c'$ and $a - b + c = a' - b' + c'$, so $b = b'$. This proves that a given quadratic form is equivalent to at most one reduced form. The next ten questions establish that every positive definite quadratic form is equivalent to a reduced form.

57 $0 \leqslant b \leqslant a \leqslant c$ implies $b^2 \leqslant ac$.
If $b^2 - 4ac = -3$, $4ac - 3 \leqslant ac$, so $ac \leqslant 1$ and $a = c = 1$. $b^2 - 4 = -3$ gives $b^2 = 1$, so the only reduced form with discriminant -3 is $x^2 + xy + y^2$.

58 If $b^2 - 4ac = -4$ and $b^2 \leqslant ac$, $4ac - 4 \leqslant ac$ so $3ac \leqslant 4$ and $a = c = 1$. Now $b^2 = 0$ and the only reduced form with discriminant -4 is $x^2 + y^2$. If $b^2 - 4ac = -12$ and $b^2 \leqslant ac$, $4ac - 12 \leqslant ac$ so $3ac \leqslant 12$ or $ac \leqslant 4$. Since $a^2 \leqslant ac$, $a = 1$ or 2.

a	c	$b^2 = 4ac - 12$	b	
1	1	−8	—	
1	2	−4	—	
1	3	0	0	$x^2 + 3y^2$
1	4	4	$2 > 1$	not reduced
2	1	not reduced		
2	2	4	2	$2x^2 + 2xy + 2y^2$

59 If $b^2 - 4ac = -d$ and $b^2 \leqslant ac$, $4ac - d \leqslant ac$ so $3ac \leqslant d$ and $ac \leqslant \tfrac{1}{3}d$. Thus $c \leqslant \tfrac{1}{3}d$ and since $0 \leqslant b \leqslant a \leqslant c$, there are at most a finite number of reduced forms with discriminant $-d$. There is a list of all reduced forms with discriminants between -3 and -83 in Davenport (1968). Davenport calls a form reduced when $0 \leqslant |b| \leqslant a \leqslant c$ because his definition of unimodular transformations insists on determinants of plus one.

60 $2(x + y)^2 - 11(x + y)y + 18y^2 = 2x^2 - 7xy + 9y^2$.

$2(x + y)^2 - 7(x + y)y + 9y^2 = 2x^2 - 3xy + 4y^2$.

$2(x + y)^2 - 3(x + y)y + 4y^2 = 2x^2 + xy + 3y^2$, reduced.

$2(x + 3y)^2 - 11(x + 3y)y + 18y^2 = 2x^2 + xy + 3y^2$.

Because $(x, y)\begin{pmatrix} 1 & 0 \\ 1 & 1 \end{pmatrix} = (x + y, y)$ and $\begin{pmatrix} 1 & 0 \\ 1 & 1 \end{pmatrix}$ is unimodular this substitution produces equivalent forms.

$(x, y)\begin{pmatrix} 1 & 0 \\ 3 & 1 \end{pmatrix} = (x + 3y, y)$, and $\begin{pmatrix} 1 & 0 \\ 3 & 1 \end{pmatrix}$ is also unimodular.

61 $2(x-y)^2+13(x-y)y+24y^2=2x^2+9xy+13y^2$.
$2(x-y)^2+9(x-y)y+13y^2=2x^2+5xy+6y^2$.
$2(x-y)^2+5(x-y)y+6y^2=2x^2+xy+3y^2$, reduced.
$2(x-3y)^2+13(x-3y)y+24y^2=2x^2+xy+3y^2$.

$(x, y)\begin{pmatrix} 1 & 0 \\ -1 & 1 \end{pmatrix}=(x-y, y)$ and $(x, y)\begin{pmatrix} 1 & 0 \\ -3 & 1 \end{pmatrix}=(x-3y, y)$.

Both $\begin{pmatrix} 1 & 0 \\ -1 & 1 \end{pmatrix}$ and $\begin{pmatrix} 1 & 0 \\ -3 & 1 \end{pmatrix}$ are unimodular matrices.

62 $3(x+ny)^2+50(x+ny)y+211y^2=3x^2+(6n+50)xy+(3n^2+50n+211)y^2$.
$6n+50$ is numerically least when $n=-8$.
$3(x-8y)^2+50(x-8y)y+211y^2=3x^2+2xy+3y^2$, reduced.
$(x, y)\begin{pmatrix} 1 & 0 \\ -8 & 1 \end{pmatrix}=(x-8y, y)$ and $\begin{pmatrix} 1 & 0 \\ -8 & 1 \end{pmatrix}$ is a unimodular matrix.

63 $3(x+ny)^2+47(x+ny)y+185y^2=3x^2+(6n+47)xy+(3n^2+47n+185)y^2$.
$6n+47$ is numerically least when $n=-8$.
$3(x-8y)^2+47(x-8y)y+185y^2=3x^2-xy+y^2$, not reduced.
By the substitution of (y, x) for (x, y)
$3x^2-xy+y^2$ is equivalent to $x^2-xy+3y^2$,
and by the substitution of $(-x, y)$ for (x, y),
$x^2-xy+3y^2$ is equivalent to $x^2+xy+3y^2$, reduced.

$(x, y)\begin{pmatrix} 1 & 0 \\ -8 & 1 \end{pmatrix}=(x-8y, y)$, $(x, y)\begin{pmatrix} 0 & 1 \\ 1 & 0 \end{pmatrix}=(y, x)$,

$(x, y)\begin{pmatrix} -1 & 0 \\ 0 & 1 \end{pmatrix}=(-x, y)$.

The matrices $\begin{pmatrix} 1 & 0 \\ -8 & 1 \end{pmatrix}$, $\begin{pmatrix} 0 & 1 \\ 1 & 0 \end{pmatrix}$ and $\begin{pmatrix} -1 & 0 \\ 0 & 1 \end{pmatrix}$ are all unimodular.

$\begin{pmatrix} -1 & 0 \\ 0 & 1 \end{pmatrix}\begin{pmatrix} 0 & 1 \\ 1 & 0 \end{pmatrix}\begin{pmatrix} 1 & 0 \\ -8 & 1 \end{pmatrix}=\begin{pmatrix} 8 & -1 \\ 1 & 0 \end{pmatrix}$, $(x, y)\begin{pmatrix} 8 & -1 \\ 1 & 0 \end{pmatrix}=(8x+y, -x)$.

$3(8x+y)^2+47(8x+y)(-x)+185(-x)^2=x^2+xy+3y^2$.

64 $3(x-8y)^2+46(x-8y)y+177y^2=3x^2-2xy+y^2$.
By the substitution of (y, x) for (x, y) this is equivalent to
$x^2-2xy+3y^2$.
$(x+y)^2-2(x+y)y+3y^2=x^2+2y^2$, reduced.

65 $a(x+ny)^2+b(x+ny)y+cy^2=ax^2+(2an+b)xy+(an^2+bn+c)y^2$ so $a=a'$. There is exactly one integer congruent to b (mod $2a$) between $-a+1$ and a, since $-a+1, -a+2, \ldots, -1, 0, 1, \ldots, a-1, a$ forms a complete set of residues modulo $2a$. Choose n so that b' is this integer.

66 As in q 65 we can transform $ax^2+bxy+cy^2$ into $ax^2+b'xy+c'y^2$ where $|b'|\leqslant a$.
By means of the substitution $(-x, y)$ for (x, y), if necessary, this is equivalent to $ax^2+|b'|xy+c'y^2$. If $a\leqslant c'$, this form is reduced. If $a>c'$, the substitution of (y, x) for (x, y) shows the form is equivalent to $c'x^2+|b'|xy+ay^2$.
If $|b'|\leqslant c'$, this form is reduced. If $|b'|>c'$, the substitution and method of q 65 will produce an equivalent form $c'x^2+b''xy+a'y^2$ with $|b''|\leqslant c'$ and, substituting $(-x, y)$ for (x, y) if necessary, an equivalent form $c'x^2+|b''|xy+a'y^2$.
If $a'\geqslant c'$, this is a reduced form. If not, the procedure above may be repeated. These procedures must terminate with a reduced form because the coefficients of x^2 and y^2 are both positive and the procedure reduces the smaller coefficient. This cannot be done indefinitely many times.

67 From q 56, each equivalence class contains at most one reduced form. From q 66, each equivalence class contains at least one reduced form. So each equivalence class contains exactly one reduced form, and the result follows from q 59.

68 See table 3.2 and q 4.42; -2 is a quadratic residue modulo 11 and modulo 17. $3^2-11\cdot 1=-2$, $7^2-17\cdot 3=-2$. Consequently $6^2-4\cdot 11\cdot 1=-8$ and $14^2-4\cdot 17\cdot 3=-8$, so required forms are $11x^2+6xy+y^2$ and $17x^2+14xy+3y^2$. Now if $ax^2+bxy+cy^2$ is a reduced form with discriminant -8, $b^2-4ac=-8$, so $b^2=4ac-8$ and $b^2\leqslant ac$ implies $ac\leqslant\frac{8}{3}$, so $a=1$, $c=1$ or 2. But $a=c=1$ implies $b^2-4=-8$ which is not possible, and $a=1$, $c=2$ implies $b=0$. So the only reduced form with discriminant -8 is x^2+2y^2. The two quadratic forms must be equivalent to this, so both 11 and 17 are representable by x^2+2y^2.

69 From q 4.42,

$$\left(\frac{-2}{p}\right)=1 \text{ when } p\equiv 1 \text{ or } 3 \pmod 8$$

and

$$\left(\frac{-2}{p}\right)=-1 \text{ when } p\equiv 5 \text{ or } 7 \pmod 8.$$

Since x^2+2y^2 is the unique reduced form with discriminant -8, p is representable by this form if there is a quadratic form $px^2+hxy+ly^2$ with discriminant -8, that is if there are integers h, l such that $h^2-4pl=-8$. For this, h must be even and then $(\frac{1}{2}h)^2-pl=-2$ so -2 must be a quadratic residue modulo p.
Conversely if -2 is a quadratic residue mod p, there are h and l satisfying $h^2-pl=-2$, and then $px^2+2hxy+l^2y$ is a quadratic form with discriminant -8. Since this is equivalent to x^2+2y^2, p is representable by this form.

70 The given equation shows that the set is closed under multiplication. Clearly any square number is representable in this form. From q 69 the primes which are representable are those congruent to 1 or 3 (mod 8) and 2 is also representable. So any integer is representable if in its prime factorisation primes congruent to 5 or 7 (mod 8) occur to an even power. If $q|x^2+2y^2$ and q is a prime congruent to 5 or 7 (mod 8) then y has a factor q, for if not there is a z such that $yz\equiv 1 \pmod q$ and then $(zx)^2+2\equiv 0 \pmod q$ and q has a quadratic residue -2. Thus primes congruent to 5 or 7 (mod 8) must occur to an even power in the factorisation of a representable number.

71 If $ax^2+bxy+cy^2$ is a reduced quadratic form with discriminant -20, then $b^2-4ac=-20$ and $b^2\leqslant ac$, so $4ac-20\leqslant ac$ and $ac\leqslant 6$. $4ac\leqslant 24$ so $b^2\leqslant 4$.
If $b=0$, $ac=5$, so $a=1$, $c=5$.
If $b=1$, $4ac=21$ which is impossible.
If $b=2$, $4ac=24$, so $ac=6$ and since $b\leqslant a\leqslant c$, $a=2$ and $c=3$.
The two reduced forms with discriminant -20 are x^2+5y^2 and $2x^2+2xy+3y^2$.
$30x^2+10xy+y^2$ has discriminant -20, so it is equivalent to one or other of the previous forms. Thus 30 is properly representable by one or other of these forms. Actually $5^2+5\cdot 1^2=30$.

72 n is properly representable by $ax^2+bxy+cy^2$ if and only if there is an equivalent quadratic form $nx^2+hxy+ly^2$. If there is only one reduced form with the discriminant b^2-4ac then every positive definite quadratic form with this discriminant is equivalent, so n is representable if there are integers h and l such that

$$h^2-4nl=b^2-4ac.$$

Such integers will exist if

$$h^2\equiv b^2-4ac \pmod{4n},$$

that is if b^2-4ac is a quadratic residue modulo $4n$.

73 $p=\pm 1$, $q=0$, $r=0$, $s=\pm 1$ corresponding to the four matrices:

$$\begin{pmatrix}1&0\\0&1\end{pmatrix}, \begin{pmatrix}-1&0\\0&-1\end{pmatrix}, \begin{pmatrix}1&0\\0&-1\end{pmatrix} \text{ and } \begin{pmatrix}-1&0\\0&1\end{pmatrix}, \text{ isomorphic to } C_2\times C_2.$$

74 Either $p=\pm 1$, $q=0$, $r=0$, $s=\pm 1$ or $p=0$, $q=\pm 1$, $r=\pm 1$, $s=0$, corresponding to the eight matrices:

$$\begin{pmatrix}1&0\\0&1\end{pmatrix}, \begin{pmatrix}-1&0\\0&-1\end{pmatrix}, \begin{pmatrix}1&0\\0&-1\end{pmatrix}, \begin{pmatrix}-1&0\\0&1\end{pmatrix}, \begin{pmatrix}0&1\\1&0\end{pmatrix},$$

$$\begin{pmatrix}0&-1\\-1&0\end{pmatrix}, \begin{pmatrix}0&1\\-1&0\end{pmatrix}, \begin{pmatrix}0&-1\\1&0\end{pmatrix},$$

isomorphic to D_4.

75 By q 55 properly represented least values of $ax^2+bxy+cy^2$ occur when $|x|, |y| \leq 1$, so $p^2+pq+q^2=1$ only when $(p, q)=(\pm 1, 0)$, $(0, \pm 1)$ or $(\pm 1, \mp 1)$.
When $(p, q)=(1, 0)$, $2r+s=1$ and $r^2+rs+s^2=1$ so $(r, s)=(0, 1)$ or $(1, -1)$.
When $(p, q)=(-1, 0)$, $-2r-s=1$, so $(r, s)=(0, -1)$ or $(-1, 1)$.
When $(p, q)=(0, 1)$, $r+2s=1$, so $(r, s)=(1, 0)$ or $(-1, 1)$.
When $(p, q)=(0, -1)$, $-r-2s=1$, so $(r, s)=(-1, 0)$ or $(1, -1)$.
When $(p, q)=(1, -1)$, $r-s=1$, so $(r, s)=(1, 0)$ or $(0, -1)$.
When $(p, q)=(-1, 1)$, $-r+s=1$, so $(r, s)=(-1, 0)$ or $(0, 1)$.
The matrix $\begin{pmatrix}1 & -1\\ 1 & 0\end{pmatrix}$ is of order 6 and with $\begin{pmatrix}0 & 1\\ 1 & 0\end{pmatrix}$, of order 2, generates the group D_6.

Fig. 8.4. Graphs of $x^2+xy+y^2=1, 3, 4$

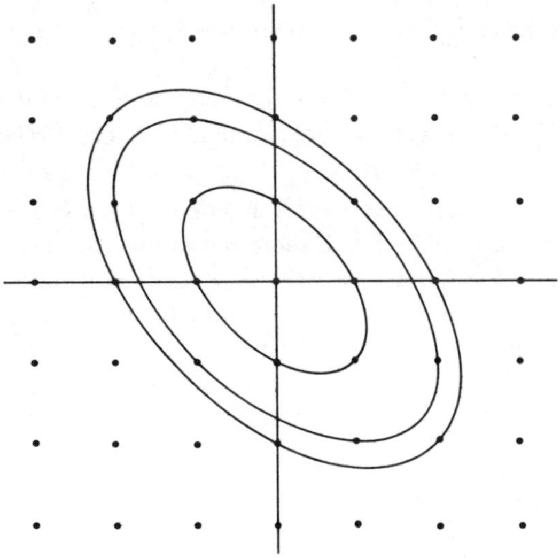

76 (i) The group of automorphs has four elements and is isomorphic to that of q 73.
(ii) The group of automorphs has two elements, the identity and $(x, y) \to (-x, -y)$.
(iii) The group of automorphs has eight elements and is isomorphic to that of q 74.
(iv) As (i).
(v) The group of automorphs has twelve elements and is isomorphic to that of q 75.

77 $(MPM^{-1})^T = M^{-1T}P^TM^T$, so $(MPM^{-1})MAM^T(MPM^{-1})^T = MPAP^TM^T$ and this equals MAM^T if and only if $PAP^T = A$.
Thus there is a one–one correspondence $P \to MPM^{-1}$ between the automorphs of the form $(x \quad y)A\begin{pmatrix} x \\ y \end{pmatrix}$ and the equivalent form $(x \quad y)MAM^T\begin{pmatrix} x \\ y \end{pmatrix}$, and this correspondence is an isomorphism.
Since every positive definite quadratic form is equivalent to a reduced form and the classification of q 76 was exhaustive for reduced forms, the classification of groups of automorphs obtained there was also exhaustive.

Historical note

In the year 1816, J. Farey and A. L. Cauchy established the result of q 14.

L. Euler made a careful examination of the quadratic forms x^2+3y^2 and x^2+2y^2 in the years 1761 and 1763 respectively. In 1773, J. L. Lagrange, using the notion of equivalence which we have adopted in q 25, showed that every positive definite form is equivalent to a reduced form and that the number of reduced forms with given discriminant is finite. He also proved our theorem (q 49 and q 50) on proper representation. In 1798, A. M. Legendre showed that distinct reduced forms could not be equivalent. In 1801, C. F. Gauss refined and extended the theory and distinguished between proper and improper equivalence. (Two quadratic forms which are equivalent under the substitution of $(px+ry, qx+sy)$ for (x, y) are said to be properly equivalent when $ps-qr=1$, and improperly equivalent when $ps-qr=-1$.) Gauss determined all automorphs under proper equivalence. A detailed history of the subject is given in chapter 1 of volume 3 of Dickson (1950).

9

Geometry of numbers

Subgroups of a square lattice

1 In fig. 9.1, a rectangular lattice has been superimposed upon a square lattice. Take a point common to the two lattices as origin, and label the points of the square lattice with the elements of $\mathbf{Z}\times\mathbf{Z}=\mathbf{Z}^2$, in the usual way. Name the points of the rectangular lattice and, supposing that it extends to infinity, show that the points of the rectangular lattice form a subgroup of $\mathbf{Z}^2$ under vector addition. Name a pair of generators for this subgroup.

Fig. 9.1

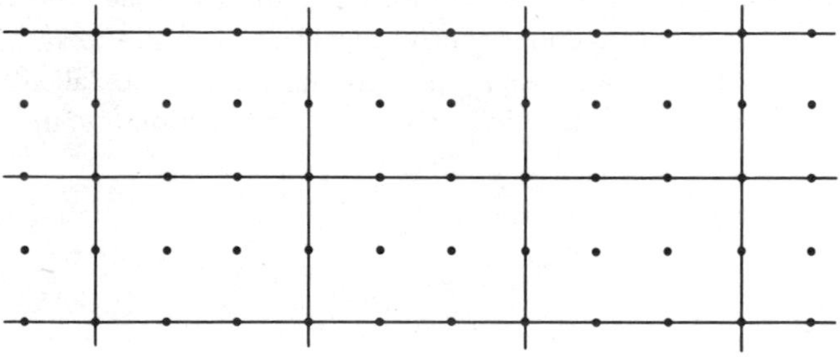

Throughout this chapter, it should be presumed that the underlying square lattice has been labelled with the set $\mathbf{Z}^2$ and that a point common to such lattices as are under discussion has been chosen as origin.

2 In fig. 9.2, find the coordinates of the points of the parallelogram lattice superimposed on $\mathbf{Z}^2$, and show that these points form a subgroup of $\mathbf{Z}^2$ under vector addition. Name a pair of generators for this subgroup.

Fig. 9.2

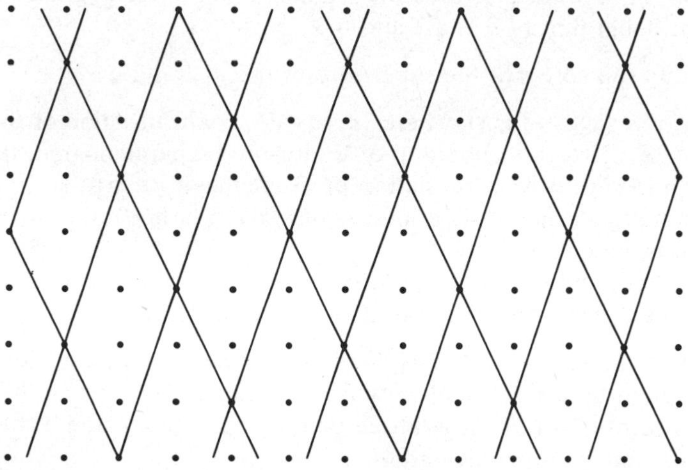

3 In fig. 9.3, find the coordinates of the parallelogram lattice superimposed on $\mathbf{Z}^2$, and show that these points form a subgroup of $\mathbf{Z}^2$ under vector addition. Name a pair of generators for this subgroup.

Fig. 9.3

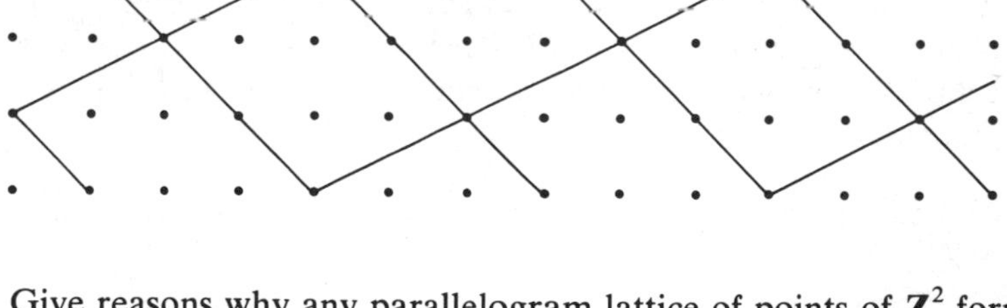

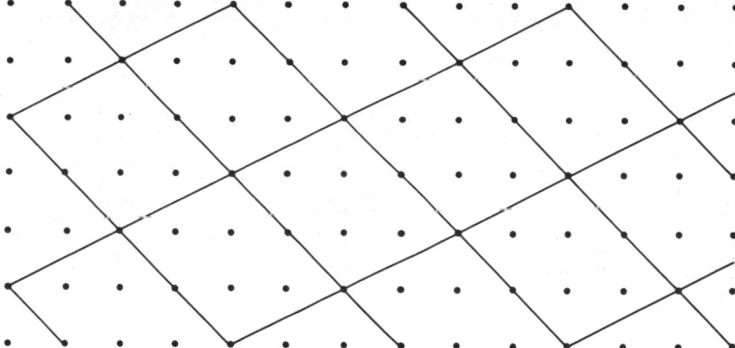

4 Give reasons why any parallelogram lattice of points of $\mathbf{Z}^2$ forms a subgroup of $\mathbf{Z}^2$, provided that the origin is common to both lattices.

5 If A and B are points of $\mathbf{Z}^2$ not collinear with the origin O, then the four points O, A, B and $A+B$ are the vertices of a

parallelogram, which we call a *fundamental parallelogram* of the subgroup $\langle A, B\rangle$. Find the area of a fundamental parallelogram of each of the subgroups of q 1, q 2 and q 3.

6 Identify the cosets of the subgroups of q 1, q 2 and q 3.

7 τ: $(x, y) \to (x, y) + (a, b)$, where $(a, b) \in \mathbf{Z}^2$, is a translation of the lattice $\mathbf{Z}^2$. If G is a subgroup of $\mathbf{Z}^2$ forming a parallelogram lattice, τ is said to be a translation of G whenever $(a, b) \in G$. Determine all the possible images of a given lattice point under the translations of G,
(i) when the given lattice point is in G,
(ii) when the given lattice point is not in G.

8 If O, A, B, $A+B$ are the vertices of a fundamental parallelogram for the group $\langle A, B\rangle$, explain why every point of $\mathbf{Z}^2$ lies in the same coset of $\langle A, B\rangle$ as a lattice point lying within or on a side of the fundamental parallelogram.

9 What is the subgroup of $\mathbf{Z}^2$ generated by $(6, 0)$ and $(21, 0)$?

10 What is the subgroup of $\mathbf{Z}^2$ generated by $(20, 10)$ and $(30, 15)$?

11 If a subgroup of $\mathbf{Z}^2$ is cyclic, what is the geometrical arrangement of the corresponding lattice points?

12 Under what conditions is the subgroup of $\mathbf{Z}^2$ generated by (a, b) and (c, d) cyclic?

13 Find two elements of $\mathbf{Z}^2$ which generate the same subgroup as $(2, 0)$, $(4, 4)$, and $(5, 2)$.

14 If G is a subgroup of $\mathbf{Z}^2$ and (a, b) and (c, d) are elements of G which are not collinear with $(0, 0)$, and the distance of these two points from $(0, 0)$ is minimal among the elements of G, prove that the group which (a, b) and (c, d) generate is a subgroup of G and if there were an element in G which was not in $\langle (a, b), (c, d)\rangle$, a contradiction would arise concerning the elements in G of minimal distance from $(0, 0)$.

15 Classify the geometrical appearance of all possible subgroups of $\mathbf{Z}^2$.

16 If we denote by Π, the region within the fundamental parallelogram O, A, B, $A+B$ and on its sides, except for the line segments $[A, A+B]$ and $[B, A+B]$, what is the set formed by the union of all images of Π under the translations of the lattice $\langle A, B\rangle$? Can two of these images overlap?

⟦193⟧

17 By reference to q 8, explain why the number of cosets of $\langle A, B\rangle$ in $\mathbf{Z}^2$ is equal to the number of lattice points of $\mathbf{Z}^2$ in Π, using the definition of Π in q 16.

18 Taking the printed squares as units, use coordinates to find the areas of the parallelograms in fig. 9.4. Can you find a connection between the areas of the parallelograms and the number of lattice points within their regions as defined in q 16?

Fig. 9.4

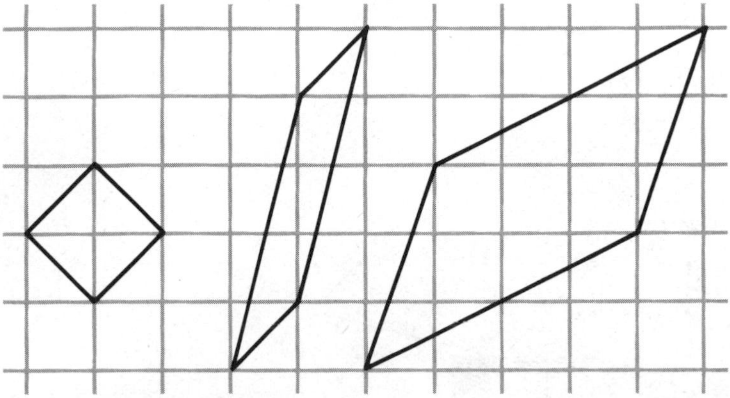

19 For each of the lattices of figs. 9.1, 9.2 and 9.3, choose a fundamental parallelogram and shade those unit squares for which the lower left hand vertex lies within the region of the fundamental parallelogram. How does the area of the shaded unit squares compare with the area of the region?

20 For each of the lattices of figs. 9.5, 9.6 and 9.7, a fundamental parallelogram has been chosen, and the unit squares with least vertex within its region have been marked. Give geometrical reasons why the area of marked squares is equal to the area of the parallelogram. (The least vertex of a square is the vertex for which both coordinates are least. The definition is unambiguous provided the sides of the squares are parallel to the axes.)

21 If O, A, B are non-collinear points of $\mathbf{Z}^2$, we define Π as in q 16 and we define a region Σ as the union of those unit squares with least vertex in Π. To prevent the squares overlapping, we exclude their two sides with greatest x and greatest y coordinates. Can any two images of Σ under the translations of the lattice $\langle A, B\rangle$ overlap?

⟦194⟧

Fig. 9.5

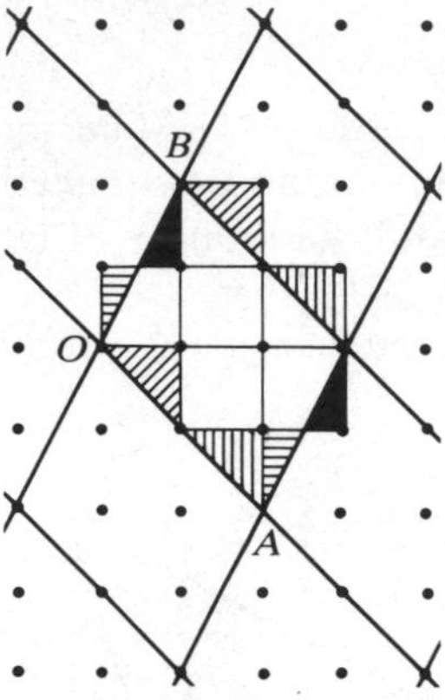

Fig. 9.6

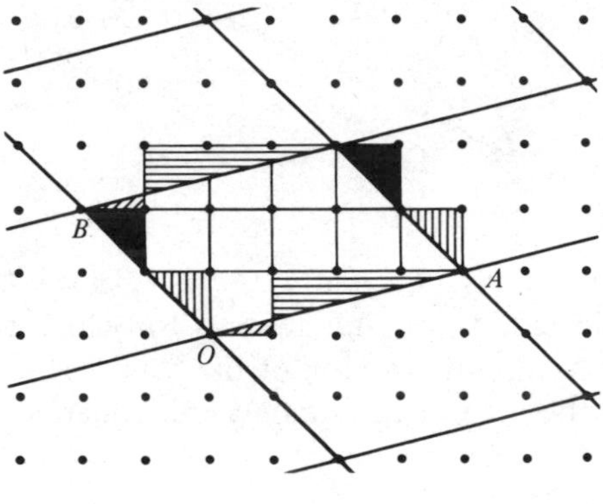

Fig. 9.7

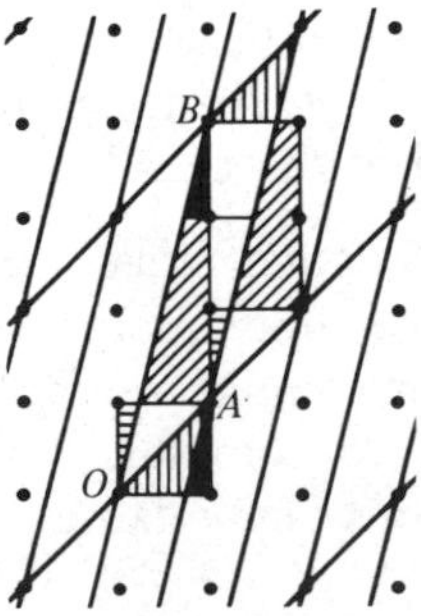

Does every point of the plane, $\mathbf{R}^2$, belong to an image of Σ under a translation of the lattice?

22 With the notation of q 21, let τ be a translation of $\langle A, B\rangle$. Why should the regions of the plane

$\tau(\Pi)\cap\Sigma$ and $\Pi\cap\tau^{-1}(\Sigma)$

be congruent, and thus have the same area?
If the union of all regions of the type $\tau(\Pi)\cap\Sigma$ is formed for all admissible τ, why is the resulting region Σ?
If the union of all regions of the type $\Pi\cap\tau^{-1}(\Sigma)$ is formed for all admissible τ, why is the resulting region Π?
Deduce that Σ and Π have the same area.

23 Prove that the area of the parallelogram O, A, B, $A+B$ is the number of cosets of $\langle A, B\rangle$ in $\mathbf{Z}^2$. This major theorem has no standard name.

24 Must any two, perhaps differently shaped, fundamental parallelograms for the same lattice $\langle P, Q\rangle$ of points have the same area?

25 Prove that the set $\{(x, y)|\ x+2y\equiv 0 \pmod 5\}$ forms a subgroup of $\mathbf{Z}^2$. Determine the number of its cosets from numerical considerations. Illustrate this subgroup on a square lattice. Find a fundamental parallelogram, and count the lattice points within its region to check your answer.

26 Prove that the set $\{(x, y)|\ ax+by\equiv 0 \pmod n\}$ forms a subgroup of $\mathbf{Z}^2$.

27 How many cosets do each of the following subgroups of $\mathbf{Z}^2$ have?
(i) $\{(x, y)|\ x+2y\equiv 0 \pmod{10}\}$,
(ii) $\{(x, y)|\ 2x+5y\equiv 0 \pmod{10}\}$,
(iii) $\{(x, y)|\ 2x+6y\equiv 0 \pmod{10}\}$.
In each case, establish that the subgroup forms a parallelogram lattice by identifying a fundamental parallelogram with the area you would expect from your enumeration of cosets.

28 Use the Chinese remainder theorem (q 2.18) to determine the area of the fundamental parallelogram of the subgroup

$\{(x, y)|\ x+y\equiv 0 \pmod 3\}\cap\{(x, y)|\ x+2y\equiv 0 \pmod 5\}$ of $\mathbf{Z}^2$.

Minkowski's theorem in two dimensions

We first develop the notions of openness, convexity and symmetry which are used in this theorem.

⟦194⟧

29 An *open region* in the plane is a region which contains a circular neighbourhood of each of its points. A circular neighbourhood of a point consists of the whole of the interior of a circle with centre at that point.

Which of the following subsets of the plane are open regions?

(i) a point,
(ii) a line segment,
(iii) the interior of a circle without its circumference,
(iv) the interior of a circle together with its circumference,
(v) the region $\{(x, y) \mid -1 < x < 1\}$,
(vi) the region $\{(x, y) \mid -1 \leq x \leq 1\}$,
(vii) the region $\{(x, y) \mid -\tfrac{1}{2} < x < \tfrac{1}{2}, -\tfrac{1}{2} < y < \tfrac{1}{2}\}$.

30 An open region, R, in the plane, has area Δ, contains the lattice point $(0, 0)$ and lies wholly within the square $\{(x, y) \mid -r < x < r, -r < y < r\}$. We denote by $R_{(a,b)}$ the image of the region R under the translation $(x, y) \to (x, y) + (a, b)$, of the integral lattice $\mathbf{Z}^2$ which maps $(0, 0)$ to (a, b). We suppose further that the regions $R_{(a,b)}$ and $R_{(c,d)}$ have no point in common unless $(a, b) = (c, d)$.

(i) Find a square which contains the nine regions $R_{(a,b)}$ for $a, b = 0, 1, -1$.
(ii) Find a square which contains the twenty-five regions $R_{(a,b)}$ for $a, b = 0, \pm 1, \pm 2$.
(iii) Find a square which contains the $(2n+1)^2$ regions $R_{(a,b)}$ for $a, b = 0, \pm 1, \pm 2, \ldots, \pm n$.
(iv) Prove that $\Delta \leq 4r^2$.
(v) Prove that $9\Delta \leq (2+2r)^2$.
(vi) Prove that $25\Delta \leq (4+2r)^2$.
(vii) Prove that $(2n+1)^2\Delta \leq (2n+2r)^2$, for all positive integers n.
(viii) Prove that

$$\Delta \leq \left(1 + \frac{r - \frac{1}{2}}{n + \frac{1}{2}}\right)^2,$$

for all positive integers n.

(ix) Prove that Δ cannot be greater than 1.
(x) Name a region R with $\Delta = 1$ which satisfies the conditions of this question.

The next four questions are concerned with the images of interiors under a linear transformation.

31 If $0 < k < 1$ and a and c are real numbers, prove that $ka + (1-k)c$ lies between a and c.

32 For any real number k, prove that the point
$(ak+(1-k)c,\ bk+(1-k)d)$ is collinear with (a, b) and (c, d).

33 Deduce from q 31 and q 32 that
$\{k(a, b)+(1-k)(c, d) \mid 0 \leqslant k \leqslant 1\}$
is the line segment joining (a, b) to (c, d).

34 If $A \to A'$ and $B \to B'$ under a linear transformation of the plane, $\mathbf{R}^2$, prove that the line segment $[A, B]$ is mapped onto the line segment $[A', B']$.
Prove also that if a triangle ABC is mapped onto a triangle $A'B'C'$ by a linear transformation of the plane, $\mathbf{R}^2$, then the interior of ABC is mapped onto the interior of $A'B'C'$ under this transformation.

35 A region is said to be *convex* if it contains the whole of the line segment joining any two of its points. Which of the following sets is convex?
(i) a point,
(ii) a line segment,
(iii) the interior of a semi-circle,
(iv) the interior of a semi-circle, together with its boundary,
(v) the boundary of a semi-circle,
(vi) the interior of a crescent.

36 A plane region is said to be *symmetrical about the point O*, if for each point A of the region, it also contains the image of A under a half-turn about O.
Is there a point about which the following regions are symmetrical?
(i) a circle,
(ii) a rectangle,
(iii) an equilateral triangle,
(iv) the infinite strip between two parallel lines.

37 Give examples of letters
(i) which are convex and symmetrical about a point,
(ii) which are convex but not symmetrical about any point,
(iii) which are symmetrical about a point but not convex,
(iv) which are neither convex nor symmetrical about any point.

38 Let R be a plane convex region which is symmetrical about the point O, and let R_A be the image of R under the translation, τ, of the plane which maps O to A. If R and R_A overlap and $P \in R \cap R_A$, explain why $\tau^{-1}(P) \in R$. Let Q be the image of $\tau^{-1}(P)$

under a half-turn about O. By considering the parallelogram with vertices P, $\tau^{-1}(P)$, Q, $\tau(Q)$, prove that the mid-point of OA lies in R. Deduce that if the image of the region R under an enlargement by a scale factor 2 with centre O is denoted by $2R$, then $2R$ contains A.

39 If R is a convex region symmetrical about $(0, 0)$ and of area greater than 4, denote by $\frac{1}{2}R$, the image of R under a shrink by a scale factor $\frac{1}{2}$ with centre $(0, 0)$. Prove that

(i) $\frac{1}{2}R$ overlaps $\tau(\frac{1}{2}R)$ for some translation τ of the integral lattice $\mathbf{Z}^2$, by q 30,

(ii) $\frac{1}{2}R$ contains the point mid-way between $(0, 0)$ and $\tau(0, 0)$, by q 38,

(iii) R contains a lattice point $\mathbf{Z}^2$ other than $(0, 0)$.

(*Minkowski's theorem*)

The next two questions provide a modest but useful generalisation of Minkowski's theorem in two dimensions.

40 Consider an arbitrary parallelogram lattice in the plane, with each fundamental parallelogram having area A. Label the vertices of one fundamental parallelogram, $(0, 0)$, $(1, 0)$, $(1, 1)$, $(0, 1)$, and extend the labelling throughout the lattice in such a way as to preserve the conventional definition of vector addition. Let R be an open region of area Δ, containing $(0, 0)$ and wholly contained within the parallelogram with sides $x = \pm r$, $y = \pm r$. Let $R_{(a,b)}$ be the image of R under the translation of this parallelogram lattice mapping $(0, 0)$, to (a, b). Suppose further, that $R_{(a,b)}$ and $R_{(c,d)}$ do not overlap unless $(a, b) = (c, d)$. Establish that

(i) $\Delta \leqslant (2r)^2 A$,

(ii) $9\Delta \leqslant (2+2r)^2 A$,

(iii) $25\Delta \leqslant (4+2r)^2 A$,

(iv) $(2n+1)^2 \Delta \leqslant (2n+2r)^2 A$, for all positive integers n,

(v) $\Delta \leqslant A(1+(r-\frac{1}{2})/(n+\frac{1}{2}))^2$;

and deduce that it is not possible for Δ to be greater than A.

41 For any parallelogram lattice with fundamental parallelogram having area A, prove that a convex region which is symmetrical about one lattice point and has area greater than $4A$, necessarily contains at least one more lattice point.

The next five questions use Minkowski's theorem to obtain alternative proofs of results about the representation of numbers by particular quadratic forms.

42 Working on a conventional square lattice, determine the area of a fundamental parallelogram of the lattice

$L=\{(x, y)|\ x-12y\equiv 0 \pmod{29}\}$.

Given that $12^2\equiv -1 \pmod{29}$, prove that $x^2+y^2\equiv 0 \pmod{29}$ for every point of the lattice L.

Show that the circle $x^2+y^2=\frac{4}{3}\cdot 29$ contains a point of L other than the origin.

Deduce that there are integers x and y such that $x^2+y^2=29$.

43 If p is a prime number and $u\not\equiv 0 \pmod{p}$, what is the area of a fundamental parallelogram of the lattice

$L=\{(x, y)|\ x-yu\equiv 0 \pmod{p}\}$?

If $p\equiv 1 \pmod{4}$ and $u^2\equiv -1 \pmod{p}$, prove that $x^2+y^2\equiv 0 \pmod{p}$ for every point of L.

Show that the circle $x^2+y^2=\frac{4}{3}p$ contains a point of L other than $(0, 0)$.

Deduce that there are integers x and y such that $x^2+y^2=p$.

44 Prove that the linear transformation

$$(x, y)\to(x, y)\begin{pmatrix} a & 0 \\ 0 & b \end{pmatrix},$$

where neither a nor $b=0$, transforms the circle $x^2+y^2=1$ into the ellipse $x^2/a^2+y^2/b^2=1$. Deduce that the area of the ellipse $x^2/a^2+y^2/b^2=1$ is πab.

45 Working on a conventional square lattice, determine the area of a fundamental parallelogram of the lattice

$L=\{(x, y)|\ x-7y\equiv 0 \pmod{17}\}$.

Given that $7^2\equiv -2 \pmod{17}$, prove that $x^2+2y^2\equiv 0 \pmod{17}$ for every point of L.

Show that the ellipse $x^2+2y^2=(\frac{4}{3}\sqrt{2})17$ contains a point of L other than the origin and deduce that there exist integers x and y such that $x^2+2y^2=17$.

46 Check from q 4.42 that $\left(\frac{-2}{p}\right)=1$ when $p\equiv 1, 3 \pmod{8}$.

If $p\equiv 1, 3 \pmod{8}$ and $u^2\equiv -2 \pmod{p}$ prove that $x^2+2y^2\equiv 0 \pmod{p}$ for each point of the lattice

$\{(x, y)|\ x-uy\equiv 0 \pmod{p}\}$.

Prove that the ellipse $x^2+2y^2=(\frac{4}{3}\sqrt{2})p$ contains at least one point of this lattice other than the origin, and deduce that there exist integers x and y such that $x^2+2y^2=p$.

⟦196⟧

From here to the end of the chapter, the questions are concerned with extending the results of q 1–q 46 to three dimensions.

Subgroups of a cubic lattice

47 Prove that each of the subsets of $\mathbf{Z}\times\mathbf{Z}\times\mathbf{Z}$,

$\{(x, y, z) | x \equiv 0 \pmod 2\}$,

$\{(x, y, z) | y \equiv 0 \pmod 3\}$,

and

$\{(x, y, z) | z \equiv 0 \pmod 5\}$,

is a subgroup of $\mathbf{Z}^3$ under vector addition.

How many cosets have each of these subgroups?

48 Verify that $\{(x, y, z) | 15x + 10y + 6z \equiv 0 \pmod{30}\}$ is the intersection of the subgroups of $\mathbf{Z}^3$ in q 47, and find the index of this subgroup.

If the elements of $\mathbf{Z}^3$ label the points of a cubic lattice in the conventional way, determine a fundamental parallelepiped for the lattice of this subgroup.

The next two questions lead to a formula for the volume of a parallelepiped. We need this in order to discuss unimodular transformations in three dimensions, and also to determine the index of subgroups of $\mathbf{Z}^3$.

49 If $O = (0, 0, 0)$, $A = (1, 2, 2)$ and $B = (0, 3, 4)$ are points of a cubic lattice, find the lengths of OA, OB and AB, and show that the cosine of the angle $AOB = \frac{14}{15}$. Use this result to show that the area of the triangle $OAB = \sqrt{29}/2$.

50 If $O = (0, 0, 0)$, $A = (a_1, a_2, a_3)$, $B = (b_1, b_2, b_3)$ and $D = (a_1 + b_1, a_2 + b_2, a_3 + b_3)$ are points of a cubic lattice, show that the cosine of angle

$$AOB = \frac{a_1b_1 + a_2b_2 + a_3b_3}{\sqrt{(a_1^2 + a_2^2 + a_3^2)(b_1^2 + b_2^2 + b_3^2)}}$$

and deduce that the area of the parallelogram $OADB$ is

$\sqrt{[(a_2b_3 - a_3b_2)^2 + (a_3b_1 - a_1b_3)^2 + (a_1b_2 - a_2b_1)^2]}$.

If $E = (a_2b_3 - a_3b_2, a_3b_1 - a_1b_3, a_1b_2 - a_2b_1)$, use your result about cosines to prove that OE is perpendicular to both OA and OB, and deduce that the perpendicular distance from the point $C = (c_1, c_2, c_3)$ to the parallelogram $OADB$ is OC times the cosine of the angle COE.

[[197]]

Show that the volume of the parallelepiped with edges OA, OB, OC and the remaining edges parallel and equal to these is
$|c_1(a_2b_3-a_3b_2)+c_2(a_3b_1-a_1b_3)+c_3(a_1b_2-a_2b_1)|$
which is also the absolute value of the determinant

$$\begin{vmatrix} a_1 & a_2 & a_3 \\ b_1 & b_2 & b_3 \\ c_1 & c_2 & c_3 \end{vmatrix}.$$

51 If the linear transformation

$$(x, y, z) \to (x, y, z)\begin{pmatrix} a_1 & a_2 & a_3 \\ b_1 & b_2 & b_3 \\ c_1 & c_2 & c_3 \end{pmatrix}$$

maps $\mathbf{Z}^3$ onto itself, by considering the images and pre-images of (1, 0, 0), (0, 1, 0) and (0, 0, 1), explain why

(i) $a_1, a_2, a_3, b_1, b_2, b_3, c_1, c_2$, and c_3 are all integers,

(ii) the determinant, Δ, of the given 3×3 matrix is a non-zero integer,

(iii) $$\frac{b_2c_3-b_3c_2}{\Delta}, \quad \frac{c_2a_3-c_3a_2}{\Delta}, \quad \frac{a_2b_3-a_3b_2}{\Delta},$$
$$\frac{b_3c_1-b_1c_3}{\Delta}, \quad \frac{c_3a_1-c_1a_3}{\Delta}, \quad \frac{a_3b_1-a_1b_3}{\Delta},$$
$$\frac{b_1c_2-b_2c_1}{\Delta}, \quad \frac{c_1a_2-c_2a_1}{\Delta} \quad \text{and} \quad \frac{a_1b_2-a_2b_1}{\Delta},$$

are all integers, and

(iv) assuming that the determinant of

$$\begin{pmatrix} b_2c_3-b_3c_2 & c_2a_3-c_3a_2 & a_2b_3-a_3b_2 \\ b_3c_1-b_1c_3 & c_3a_1-c_1a_3 & a_3b_1-a_1b_3 \\ b_1c_2-b_2c_1 & c_1a_2-c_2a_1 & a_1b_2-a_2b_1 \end{pmatrix}$$

is Δ^2, prove that $1/\Delta$ is an integer and deduce that $\Delta=\pm1$.

52 If A is a 3×3 matrix with integral entries, and the determinant of A is ±1, prove that the linear transformation

$(x, y, z) \to (x, y, z)A$

maps $\mathbf{Z}^3$ onto $\mathbf{Z}^3$. Such a matrix will be called a *unimodular matrix* and such a transformation a *unimodular transformation.*

The next fourteen questions are concerned with exploring the subgroups of $\mathbf{Z}^3$, particularly those of finite index.

⟦197⟧

53 If the elements of a subgroup of $\mathbf{Z}^3$ all lie in a plane through the origin, what are the possible geometric arrangements for the corresponding points?

54 If the three points A, B and C do not lie on a plane through the origin and are points of minimal distance from the origin in a subgroup G of $\mathbf{Z}^3$, explain why $G = \langle A, B, C\rangle$.

55 If a subgroup of $\mathbf{Z}^3$ does not lie wholly in one plane through the origin, what is the geometric arrangement of its corresponding points?

56 Find a set of three generators for each of the following subgroups of $\mathbf{Z}^3$ under vector addition.
(i) $\{(x, y, z) | x \equiv 0 \pmod 2\}$,
(ii) $\{(x, y, z) | y \equiv 0 \pmod 3\}$,
(iii) $\{(x, y, z) | z \equiv 0 \pmod 5\}$,
(iv) $\{(x, y, z) | x \equiv 0 \pmod 2 \text{ and } y \equiv 0 \pmod 3\}$,
(v) $\{(x, y, z) | 3x + 2y \equiv 0 \pmod 6\}$,
(vi) $\{(x, y, z) | x \equiv 0 \pmod 2,\ y \equiv 0 \pmod 3 \text{ and } z \equiv 0 \pmod 5\}$,
(vii) $\{(x, y, z) | 15x + 10y + 6z \equiv 0 \pmod{30}\}$.

57 When A, B, C are points which are not coplanar with the origin O, then the parallelepiped, with vertices O, A, B, C, $B+C$, $C+A$, $A+B$, $A+B+C$ is called a *fundamental parallelepiped* for the subgroup $\langle A, B, C\rangle$ of $\mathbf{Z}^3$.
Determine the volumes of fundamental parallelepipeds for the subgroups of q 56.

58 How many cosets does each of the subgroups of $\mathbf{Z}^3$ in q 56 have?

59 If G is a subgroup of $\mathbf{Z}^3$, define a translation of the lattice G by analogy with q 7.

60 If G is a subgroup of $\mathbf{Z}^3$ which does not lie wholly in one plane, must every point of $\mathbf{Z}^3$ lie in the same coset of G as a point in a given fundamental parallelepiped of G?

61 If we denoted by Π, the region within the fundamental parallelepiped O, A, B, C, $B+C$, $C+A$, $A+B$, $A+B+C$ and on its faces and edges, excepting the points on the three faces

A, $A+B$, $A+B+C$, $C+A$,

B, $B+C$, $A+B+C$, $A+B$,

C, $C+A$, $A+B+C$, $B+C$

and their edges, what is the set formed by the union of all images of Π under the translations of the lattice $\langle A, B, C\rangle$? Can any two images overlap?

[[198]]

62 Count the lattice points of $\mathbf{Z}^3$ in Π as defined in q 61 for the fundamental parallelepiped of the last subgroup of q 56.

63 Why must the number of cosets of $\langle A, B, C\rangle$ in $\mathbf{Z}^3$ equal the number of lattice points in Π as defined in q 61?

64 If O, A, B, C are non-coplanar points of $\mathbf{Z}^3$ we define Π as in q 61 and we define a region V as the union of those unit cubes with least vertex in Π. The least vertex of a cube is that for which all three coordinates are least. The definition is unambiguous for cubes with edges parallel to the axes. To prevent the cubes overlapping, we exclude the three faces with greatest x, greatest y and greatest z respectively from the cubes whose union forms V.
Can two images of V under the translations of the lattice $\langle A, B, C\rangle$ overlap?
Does every point of space, $\mathbf{R}^3$, lie in an image of V under a translation of the lattice?

65 With the notation of q 64, let τ be a translation of the lattice $\langle A, B, C\rangle$. Why should the regions of space $\tau(\Pi)\cap V$ and $\Pi\cap\tau^{-1}(V)$ be congruent, and thus have the same volume?
If the union of all regions of the type $\tau(\Pi)\cap V$ is formed for all admissible τ, why is the resulting region V?
If the union of all regions of the type $\Pi\cap\tau^{-1}(V)$ is formed for all admissible τ, why is the resulting region Π?
Deduce that V and Π have the same volume.

66 If O, A, B, C are not coplanar, prove that the volume of the parallelepiped $O, A, B, C, B+C, C+A, A+B, A+B+C$ is equal to the number of cosets of $\langle A, B, C\rangle$ in $\mathbf{Z}^3$. This major theorem has no standard name.

Minkowski's theorem in three dimensions

67 By analogy with q 29, define an open region in three-dimensional space. Which of the following subsets of $\mathbf{R}^3$ are open regions?
(i) a point,
(ii) a line segment,
(iii) a circular disc,
(iv) the interior of a sphere without its surface,
(v) the interior of a sphere together with its surface,
(vi) the region $\{(x, y, z) \mid -1<x<1\}$,
(vii) the region $\{(x, y, z) \mid -1\leq x\leq 1\}$,
(viii) the region $\{(x, y, z) \mid -\tfrac{1}{2}<x<\tfrac{1}{2}, -\tfrac{1}{2}<y<\tfrac{1}{2}, -\tfrac{1}{2}<z<\tfrac{1}{2}\}$.

⟦198⟧

68 Formulate a three-dimensional analogue of q 30, for an open region R containing (0, 0, 0) and such that no two of its images under the translations of $\mathbf{Z}^3$ overlap.

69 The mapping of space given by $(x, y, z) \rightarrow (-x, -y, -z)$ is called an inversion in (0, 0, 0). A region is said to be *symmetrical about the point O in space* if for each point A of the region, it also contains the image of A under an inversion in O.
Give examples of regions of $\mathbf{R}^3$ with non-zero volume which are
(i) convex and symmetrical about a point,
(ii) convex but not symmetrical about any point,
(iii) symmetrical about a point but not convex,
(iv) neither convex nor symmetrical about any point.

70 State and prove an analogue of q 38 in three-dimensional space.

71 State the three-dimensional analogue of Minkowski's theorem (q 39).

72 Establish by analogy with q 40 that if R is an open region of three-dimensional space containing (0, 0, 0), and for any parallelepiped lattice of $\mathbf{Z}^3$ no two images of R under the translations of lattice overlap, then the volume of R cannot exceed the volume of the fundamental parallelepiped.

73 Establish by analogy with q 41, for any parallelepiped lattice with fundamental parallelepiped having volume V, that a convex region, symmetrical about one lattice point and having volume greater than $8V$, necessarily contains at least one more lattice point.

Legendre's theorem on $ax^2+by^2+cz^2=0$

We now start to elaborate the techniques which are used in the proof of Legendre's theorem and this occupies the rest of the chapter.

74 If $y \equiv z \pmod 3$, prove that $15x^2+14y^2-71z^2 \equiv 0 \pmod 3$.
If $z \equiv 2y \pmod 5$, prove that $15x^2+14y^2-71z^2 \equiv 0 \pmod 5$.
Deduce that if $y \equiv z \pmod 3$ and $z \equiv 2y \pmod 5$ then $15x^2+14y^2-71z^2 \equiv 0 \pmod{15}$.
What is the volume of the fundamental parallelepiped of the lattice

$$\{(x, y, z) \mid y \equiv z \pmod 3\} \cap \{(x, y, z) \mid z \equiv 2y \pmod 5\}?$$

75 If $x \equiv z \pmod 7$, prove that $15x^2+14y^2-71z^2 \equiv 0 \pmod 7$.
If $y \equiv 2x \pmod{71}$, prove that $15x^2+14y^2-71z^2 \equiv 0 \pmod{71}$.
Deduce that if $x \equiv z \pmod 7$ and $y \equiv 2x \pmod{71}$ then $15x^2+14y^2-71z^2 \equiv 0 \pmod{497}$.

What is the volume of the fundamental parallelepiped of the lattice

$\{(x, y, z) \mid x \equiv z \pmod 7\} \cap \{(x, y, z) \mid y \equiv 2x \pmod{71}\}$?

76 If $x \equiv z \pmod 4$ and $y \equiv 0 \pmod 2$ prove that $15x^2+14y^2-71z^2 \equiv 0 \pmod 8$.

77 What is the volume of the fundamental parallelepiped of the lattice $\{(x, y, z) \mid y \equiv z \pmod 3,\ z \equiv 2y \pmod 5,\ x \equiv z \pmod 7,\ y \equiv 2x \pmod 7,\ x \equiv z \pmod 4 \text{ and } y \equiv 0 \pmod 2\}$?
Prove that each point of this lattice satisfies $15x^2+14y^2-71z^2 \equiv 0 \pmod{4 \cdot 15 \cdot 14 \cdot 71}$.

78 Prove that the linear transformation

$$(x, y, z) \to (x, y, z)\begin{pmatrix} a & 0 & 0 \\ 0 & b & 0 \\ 0 & 0 & c \end{pmatrix},$$

where $a, b, c \neq 0$, maps the sphere $x^2+y^2+z^2=1$ onto the ellipsoid $x^2/a^2+y^2/b^2+z^2/c^2=1$. Deduce that the volume of the ellipsoid $x^2/a^2+y^2/b^2+z^2/c^2=1$ is $\frac{4}{3}\pi abc$.

79 Show that the volume of the ellipsoid

$$15x^2+14y^2+71z^2=4 \cdot 15 \cdot 14 \cdot 71$$

is $\frac{4}{3}\pi\sqrt{(4 \cdot 14 \cdot 71 \cdot 4 \cdot 15 \cdot 71 \cdot 4 \cdot 15 \cdot 14)} = \frac{4}{3}\pi \cdot 8 \cdot 15 \cdot 14 \cdot 71 > 8(4 \cdot 15 \cdot 14 \cdot 71)$.

80 Deduce from q 73 and q 79 that there is a point of the lattice of q 77 other than the origin inside the ellipsoid $15x^2+14y^2+71z^2=4 \cdot 15 \cdot 14 \cdot 71$. If this lattice point, other than the origin, is (x_1, y_1, z_1), show that

$$|15x_1^2+14y_1^2-71z_1^2| < 4 \cdot 15 \cdot 14 \cdot 71$$

and deduce that

$$15x_1^2+14y_1^2-71z_1^2=0.$$

81 There exist integers x, y and z with

$$3x^2+5y^2-23z^2=0 \quad \text{and} \quad \gcd(x, y, z)=1.$$

How many of x, y and z may be even and how many odd?

82 There exist integers x, y and z with

$$3x^2+5y^2-62z^2=0 \quad \text{and} \quad \gcd(x, y, z)=1.$$

Explain why x and y must be odd, deduce that $x^2 \equiv y^2 \equiv 1 \pmod 8$ and show that z is even. Show also that $x \equiv \pm y \pmod 4$.

⟦199⟧

83 There exist integers x, y and z with

$$3x^2+7y^2-2z^2=0 \quad \text{and} \quad \gcd(x, y, z)=1.$$

Prove that $x^2 \equiv y^2 \equiv 1 \pmod 8$. Deduce that z is odd and that $x \equiv \pm y \pmod 4$.

84 If x_1, y_1 and z_1 are integers such that

$$ax_1^2+by_1^2+cz_1^2=0 \pmod 8,$$

$\gcd(ax_1, by_1, cz_1, 4abc)=1$ and c is an even number, show that

(i) both ax_1 and by_1 are odd numbers,
(ii) if z_1 is even, then $a+b\equiv 0 \pmod 8$,
(iii) if z_1 is odd, then $a+b+c\equiv 0 \pmod 8$.

Deduce that if $x\equiv y \pmod 4$ and $z\equiv z_1x \pmod 2$ then

$$ax^2+by^2+cz^2\equiv 0 \pmod 8.$$

85 If x_1, y_1 and z_1 are integers such that

$$ax_1^2+by_1^2+cz_1^2\equiv 0 \pmod p,$$

where p is an odd prime divisor of a and $\gcd(ax_1, by_1, cz_1, 4abc)=1$, show how to find an l such that

$$ax^2+by^2+cz^2\equiv 0 \pmod p,$$

when $y\equiv lz \pmod p$.

86 If x_1, y_1 and z_1 are integers such that

$$ax_1^2+by_1^2+cz_1^2\equiv 0 \pmod{4abc},$$

where a, b and c are without squares in their prime factorisation, are coprime in pairs, and $\gcd(ax_1, by_1, cz_1, 4abc)=1$, construct a lattice with fundamental parallelepiped having volume $|4abc|$ such that

$$ax^2+by^2+cz^2\equiv 0 \pmod{4abc}$$

for every point (x, y, z) of the lattice.
Determine the volume of the ellipsoid

$$|a|x^2+|b|y^2+|c|z^2=|4abc|,$$

and prove that at least one point of the lattice other than the origin lies inside the ellipsoid.
Deduce that an integral solution to

$$ax^2+by^2+cz^2=0 \quad \text{exists.}$$

(*Legendre's theorem*)

Notes and answers

For concurrent reading see chapter 3 of Hardy & Wright (1980).

1 $(3m, 2n)$. The group is generated by $(3, 0)$ and $(0, 2)$.

2 Closed under addition because of the parallelogram law of vector addition. Each element of parallelogram lattice is of the form $m(1, -2)+n(1, 3)$.

3 Each point of the parallelogram lattice has the form $m(2, -2)+n(2, 1)$.

4 Closed by the parallelogram law of addition. It contains inverses because it is symmetrical about any point common to both lattices.

5 q 1, 6; q 2, 5; q 3, 6.

6 For q 1, $(3m, 2n)+(1, 0)$,
$(3m, 2n)+(2, 0)$,
$(3m, 2n)+(0, 1)$,
$(3m, 2n)+(1, 1)$,
$(3m, 2n)+(2, 1)$,
and the subgroup itself.

For q 2, $(m+n, -2m+3n)+(1, -1)$,
$(m+n, -2m+3n)+(1, 0)$,
$(m+n, -2m+3n)+(1, 1)$,
$(m+n, -2m+3n)+(1, 2)$,
and the subgroup itself.

For q 3, $(2m+2n, -2m+n)+(1, -1)$,
$(2m+2n, -2m+n)+(2, -1)$,
$(2m+2n, -2m+n)+(3, -1)$,
$(2m+2n, -2m+n)+(1, 0)$,
$(2m+2n, -2m+n)+(2, 0)$
and the subgroup itself.

7 (i) The translations of G map G onto G.
(ii) The images of a point under the translations of G form a coset of G in $\mathbf{Z}^2$.

8 Each lattice point of $\mathbf{Z}^2$ lies in or on some parallelogram of the lattice $\langle A, B\rangle$; but this parallelogram is the image of O, A, B, $A+B$ under a translation of $\langle A, B\rangle$, so the original lattice point has a pre-image in O, A, B, $A+B$ and so belongs to the same coset.

9 $(3n, 0)$.

10 $(10n, 5n)$.

11 All the lattice points lie on a line through the origin.

12 When $(0, 0)$, (a, b), and (c, d) are collinear.

13 Use squared paper and find the points generated closest to $(0, 0)$. Any two of $(2, 0)$, $(1, 2)$, $(-1, 2)$.

14 Certainly (a, b) and (c, d) generate a parallelogram lattice. If there is a point of G which is not on this lattice, then a translation of the lattice establishes a point of G either within the triangle $(0, 0)$, (a, b), (c, d) or

within the triangle $(0, 0)$, $(-a, -b)$, $(-c, -d)$, and in either case there is a point of G nearer to $(0, 0)$ than either (a, b) or (c, d).

15 Either one point, or a line of points (one generator) or a parallelogram lattice (two generators).

16 The whole plane. If two images of Π overlap, they are identical.

17 There is no translation of $\langle A, B\rangle$ which maps one lattice point in Π to another, so each of the lattice points in Π belong to a different coset of $\langle A, B\rangle$. Under a translation of $\langle A, B\rangle$ each point of $\mathbf{Z}^2$ may be mapped to a lattice point of Π, so each coset of $\langle A, B\rangle$ has one representative in Π.

18 From q 8.5, the areas are the determinants of

$$\begin{pmatrix} 1 & -1 \\ 1 & 1 \end{pmatrix}, \quad \begin{pmatrix} 1 & 1 \\ 2 & 5 \end{pmatrix} \quad \text{and} \quad \begin{pmatrix} 4 & 2 \\ 1 & 3 \end{pmatrix}, \quad \text{that is 2, 3, and 10 respectively.}$$

The area equals the number of lattice points within, taking account of edges as in q 16. Pick's theorem (see *Introduction to geometry* by H. S. M. Coxeter, Wiley 1969) would establish the link, but we develop a more general argument in q 19–23 which may be extended to three or more dimensions.

19 The area of the shaded unit squares equals the area of the parallelogram in each case.

20 Translations of the lattice match the non-overlapping parts of the squares outside the parallelogram, with the uncovered portions of the parallelogram.

21 No two images of Σ overlap unless they are identical. Since each point of the plane belongs to a unit square and each unit square is the image of a unit square in Σ under a translation of the lattice, the whole plane is covered by the images of Σ under the translations of the lattice.

22 Under τ, $\Pi \cap \tau^{-1}(\Sigma)$ is mapped onto $\tau(\Pi) \cap \Sigma$, so these regions are congruent. Every point of $\tau(\Pi) \cap \Sigma$ lies in Σ. The set of all $\tau(\Pi)$ covers the whole plane, so the set of $\tau(\Pi) \cap \Sigma$ covers the whole of Σ. Every point of $\Pi \cap \tau^{-1}(\Sigma)$ lies in Π. The set of all $\tau^{-1}(\Sigma)$ covers the whole plane, so the set of $\Pi \cap \tau^{-1}(\Sigma)$ covers the whole of Π.

23 Number of cosets = number of lattice points in the region of the fundamental parallelogram
= number of unit squares with least vertex a lattice point in the region
= area of the unit squares
= area of the parallelogram.

24 Yes, because the number of cosets is a property of the subgroup, not a property of the choice of generators.

25 Cosets are $x+2y\equiv 0 \pmod 5$,
$x+2y\equiv 1 \pmod 5$,
$x+2y\equiv 2 \pmod 5$,
$x+2y\equiv 3 \pmod 5$
and $x+2y\equiv 4 \pmod 5$.

26 If $ax+by\equiv 0 \pmod n$ and $ax'+by'\equiv 0 \pmod n$ then $a(x+x')+b(y+y')\equiv 0 \pmod n$, so the set is closed. Also $a(-x)+b(-y)\equiv 0 \pmod n$, so it contains the inverses of all its elements.

27 (i) 10, (2, −1) and (2, 4), for example.
(ii) 10, (5, 0) and (0, 2), for example.
(iii) 5, (2, 1) and (5, 0), for example.

28 Since gcd (3, 5) = 1, for each x there is a unique y (mod 15) such that

$$x+y\equiv 0 \pmod 3$$

and

$$x+2y\equiv 0 \pmod 5.$$

The group generated by (1, 2) and (−1, 13), for example; area 15.

29 Only (iii), (v) and (vii).

30 (i) $|x|, |y|<r+1$.
(ii) $|x|, |y|<r+2$.
(iii) $|x|, |y|<r+n$.
(iv)–(vii) area of region $\leqslant$ area of square in which it lies.
(viii) follows from (vii) on division by $(2n+1)^2$.
(ix) Let $n\to\infty$ in (viii).
(x) $|x|, |y|<\frac{1}{2}$.

31 Suppose $a>c$; then $a-c>0$
so $1>k>0$
implies $a-c>k(a-c)>0$
and so $a>k(a-c)+c>c$
or $a>ka+(1-k)c>c$.
Similarly if $c>a$, since $1>1-k>0$.

32 $(ak+(1-k)c,\ bk+(1-k)d)=(c, d)+k(a-c, b-d)$.
Now the line joining the origin to $(a-c, b-d)$ is parallel to the line joining (a, b) to (c, d). Thus the point in question lies on the line through (c, d) in the direction of (a, b).

33 By q 32, each point in this set lies on the line joining (a, b) to (c, d). By q 31 each point lies on the line segment joining the two, and divides them in the positive ratio $1-k:k$.

34 If $(a, b) \to (a, b)'$ and $(c, d) \to (c, d)'$ under a linear transformation of the plane, then

$k(a, b) + (1-k)(c, d) \to k(a, b)' + (1-k)(c, d)'$,

so line segments are mapped to line segments.

Hence the sides of a triangle are mapped to the sides of a triangle, and each interior point lies on a line segment joining two points on the sides.

35 All except (v) and (vi).

36 A unique point for (i) and (ii). No point for (iii). Many points for (iv).

37 (i) Interior of O. (ii) Interior of D. (iii) Lines of H. (iv) Lines of L.

38 $P \in R_A \Rightarrow \tau^{-1}(P) \in \tau^{-1}(R_A) = R$.

Since R is symmetrical about O, $Q \in R$ and so $\tau(Q) \in R_A$. O is the mid-point of $\tau^{-1}(P)Q$ and A is the mid-point of $P\tau(Q)$, so the mid-point of OA is at the centre of the parallelogram. $P, Q \in R$ and since R is convex the mid-point of $PQ \in R$. This point is also the mid-point of OA.

39 (i) If $\frac{1}{2}R$ did not overlap any $\tau(\frac{1}{2}R)$, then by q 30, the area of $\frac{1}{2}R$ could not exceed 1.

(ii) Since $\frac{1}{2}R$ overlaps $\tau(\frac{1}{2}R)$, q 38 shows that the mid-point of (0, 0) and $\tau(0, 0)$ is in $\frac{1}{2}R$.

(iii) Thus $\tau(0, 0)$ is in R.

40 Argue as in q 30.

41 Let R be a convex region symmetrical about a lattice point and with area $>4A$. Then $\frac{1}{2}R$ has area $>A$, so by q 40 it overlaps one of its images under translations of the lattice. By the argument of q 38, $\frac{1}{2}R$ contains the mid-point of the line joining the centre of $\frac{1}{2}R$ to the centre of its relevant translation image. Thus R contains the centre of one of its images under translations of the lattice.

42 Area 29.

If $x \equiv 12y \pmod{29}$, $x^2 + y^2 \equiv 12^2y^2 + y^2 \equiv 0 \pmod{29}$. The circle $x^2 + y^2 = \frac{4}{3} \cdot 29$ is convex, symmetrical about (0, 0) and has area $\pi \cdot \frac{4}{3} \cdot 29 > 4 \cdot 29$. By q 41 it contains a point of the given lattice other than (0, 0). Let this point be (a, b). Then $a^2 + b^2 \equiv 0 \pmod{29}$ and $a^2 + b^2 < \frac{4}{3} \cdot 29 < 2 \cdot 29$, so $a^2 + b^2 = 29$.

43 Area p.

$x^2 + y^2 \equiv y^2u^2 + y^2 \equiv y^2(u^2 + 1) \equiv 0 \pmod{p}$.

The circle $x^2 + y^2 = \frac{4}{3}p$ is convex and symmetrical about (0, 0). Its area is $\pi \cdot \frac{4}{3}p > 4p$, so by q 41 it contains a point of the given lattice other than (0, 0). Let this point be (a, b), then $a^2 + b^2 \equiv 0 \pmod{p}$ and $a^2 + b^2 < \frac{4}{3}p < 2p$, so $a^2 + b^2 = p$. This is an alternative proof, independent of chapter 6, that every prime congruent to 1 (mod 4) is expressible as the sum of two squares.

44 (x, y) lies on $x^2+y^2=1$ if and only if (ax, by) lies on $x^2/a^2+y^2/b^2=1$. Under a linear transformation of the plane, areas are multiplied by the determinant of the corresponding matrix. See q 8.3 and q 8.5.

45 Area 17.

$x^2+2y^2 \equiv 49y^2+2y^2 \equiv 51y^2 \equiv 0 \pmod{17}$.

The ellipse $x^2+2y^2=(\frac{4}{3}\sqrt{2})17$ is convex and symmetrical about (0, 0). Its area is $\pi \cdot \frac{4}{3} \cdot 17 > 4 \cdot 17$, so by q 41 it contains a point of the given lattice other than (0, 0). Let this point be (a, b), then $a^2+2b^2 \equiv 0 \pmod{17}$ and $a^2+2b^2 < (\frac{4}{3}\sqrt{2})17 < 2 \cdot 17$, so $a^2+2b^2=17$.

46 Area p.

$x^2+2y^2 \equiv u^2y^2+2y^2 \equiv (u^2+2)y^2 \equiv 0 \pmod{p}$.

The ellipse is convex, has centre at the origin and area $\pi \cdot \frac{4}{3} \cdot p > 4p$. By q 41 it contains a point of the given lattice other than (0, 0). Let this point be (a, b), then $a^2+2b^2 \equiv 0 \pmod{17}$ and $a^2+2b^2 < (\frac{4}{3}\sqrt{2})17 < 2 \cdot 17$, so $a^2+2b^2=p$.

47 Each is closed and contains inverses of its elements.

Number of cosets = 2, 3, 5 respectively.

48 $x \equiv 0 \pmod 2$ and $y \equiv 0 \pmod 3 \Leftrightarrow 3x+2y \equiv 0 \pmod 6$; $3x+2y \equiv 0 \pmod 6$ and $z \equiv 0 \pmod 5 \Leftrightarrow 5(3x+2y)+6z \equiv 0 \pmod{30}$. This subgroup has 30 cosets. Cuboid bounded by $x=0, 2$; $y=0, 3$; $z=0, 5$.

49 $OA=\sqrt{(1^2+2^2+2^2)}=3$, $OB=\sqrt{(0^2+3^2+4^2)}=5$,

$AB=\sqrt{[1^2+(3-2)^2+(4-2)^2]}=\sqrt{6}$.

$AB^2=OA^2+OB^2-2 \cdot OA \cdot OB \cdot \cos AOB$

so $6=9+25-2 \cdot 5 \cdot 3 \cdot \cos AOB$ and $\cos AOB=\frac{14}{15}$.

$$\begin{aligned}\text{Area of } \Delta OAB &= \tfrac{1}{2} OA \cdot OB \sin AOB\\ &= \tfrac{1}{2} \cdot 3 \cdot 5 \cdot \sqrt{[1-(\tfrac{14}{15})^2]}\\ &= \tfrac{1}{2}\sqrt{(225-196)} = \tfrac{1}{2}\sqrt{29}.\end{aligned}$$

50 From the cosine formula, $\cos AOB = \dfrac{OA^2+OB^2-AB^2}{2 \cdot OA \cdot OB}$.

Area of parallelogram $= OA \cdot OB \sin AOB$.

OA is perpendicular to OB when $\cos AOB=0$ or $a_1b_1+a_2b_2+a_3b_3=0$.

Volume of parallelepiped = area of $OADB$ · perpendicular distance from C to this plane.

51 (i) The image of $(1, 0, 0)$ is (a_1, a_2, a_3), etc. so the entries in the matrix must be integers.

(ii) From (i), Δ is an integer. If Δ were 0, (a_1, a_2, a_3), (b_1, b_2, b_3) and (c_1, c_2, c_3) would be coplanar with the origin, and so the transformation would not map $\mathbf{Z}^3$ onto itself.

(iii) The triples here give the coordinates of the points which are mapped onto (1, 0, 0), (0, 1, 0) and (0, 0, 1) respectively.

(iv) The determinant of the matrix formed by the pre-images of (1, 0, 0), (0, 1, 0) and (0, 0, 1) is $\Delta^2/\Delta^3 = 1/\Delta$. Since the entries in this matrix are all integers, this is an integer. If both Δ and $1/\Delta$ are integers, $\Delta = \pm 1$.

52 Since A has integral entries, the transformation maps $\mathbf{Z}^3$ into $\mathbf{Z}^3$. Since the determinant of $A = \pm 1$, the pre-images of (1, 0, 0), (0, 1, 0) and (0, 0, 1) are also points of $\mathbf{Z}^3$. Thus the mapping of $\mathbf{Z}^3$ is one–one and onto $\mathbf{Z}^3$.

53 Either they lie on a line through the origin, or they form a parallelogram lattice.

54 Argument analogous to q 14.

55 A parallelepiped lattice.

56 (i) (2, 0, 0), (0, 1, 0), (0, 0, 1).
(ii) (1, 0, 0), (0, 3, 0), (0, 0, 1).
(iii) (1, 0, 0), (0, 1, 0), (0, 0, 5).
(iv) (2, 0, 0), (0, 3, 0), (0, 0, 1).
(v) as (iv).
(vi) (2, 0, 0), (0, 3, 0), (0, 0, 5).
(vii) as (vi).

57 (i) 2, (ii) 3, (iii) 5, (iv) and (v) 6, (vi) and (vii) 30.

58 Same answers as q 57.

59 $(x, y, z) \to (x, y, z) + (a, b, c)$ is a translation of G when $(a, b, c) \in G$.

60 Each point lies in some parallelepiped of the lattice, and so can be mapped into a point of the fundamental parallelepiped by a translation of the lattice.

61 Three-dimensional space. If two images overlap, they coincide.

62 $x = 0, 1$; $y = 0, 1, 2$; $z = 0, 1, 2, 3, 4$. 30 points.

63 Argue as in n 17.

64 If two images overlap, they coincide. Yes.

65 The translation τ maps $\Pi \cap \tau^{-1}(V)$ onto $\tau(\Pi) \cap V$.
The set of all $\tau(\Pi)$ fill space.
The set of all $\tau^{-1}(V)$ fill space.

66 The number of cosets of $\langle A, B, C\rangle$
= number of lattice points in the fundamental region
= number of unit cubes with least vertex in the fundamental region
= volume of these unit cubes
= volume of parallelepiped.

67 An open region in three-dimensional space is one that contains a spherical neighbourhood of each of its points.
(iv), (vi) and (viii).

68 If an open region R of three-dimensional space has volume V and contains the point $(0, 0, 0)$, and none of its images under translations of $\mathbf{Z}^3$ overlap, then if it lies within the cube $|x|, |y|, |z| < r$, $V \leq 1$.

69 (i) sphere, (ii) hemisphere, (iii) a pair of touching spheres with equal radii, (iv) a pair of touching spheres with unequal radii.

70 If R is a convex region in space which is symmetrical about a point O, and R overlaps its image under the translation which maps O to A, then the enlargement of R with centre O and scale factor 2 contains A.

71 If R is a convex region in space symmetrical about $(0, 0, 0)$ and of volume greater than 8, R contains a lattice point of $\mathbf{Z}^3$ other than $(0, 0, 0)$.

74 $15x^2+14y^2-71z^2 \equiv 14y^2-71z^2 \pmod 3$
and if $y \equiv z \pmod 3$, $14y^2-71z^2 \equiv -57z^2 \equiv 0 \pmod 3$.
$15x^2+14y^2-71z^2 \equiv 14y^2-71z^2 \pmod 5$
and if $z \equiv 2y \pmod 5$, $14y^2-71z^2 \equiv 14y^2-284y^2 \equiv -270y^2 \equiv 0 \pmod 5$.
Volume 15.

75 $15x^2+14y^2-71z^2 \equiv 15x^2-71z^2 \pmod 7$
and if $x \equiv z \pmod 7$, $15x^2-71z^2 \equiv -56z^2 \equiv 0 \pmod 7$.
$15x^2+14y-71z^2 \equiv 15x^2+14y^2 \pmod{71}$,
and if $y \equiv 2x \pmod{71}$, $15x^2+14y^2 \equiv 71x^2 \equiv 0 \pmod{71}$.
Volume 497.

76 $y \equiv 0 \pmod 2 \Rightarrow y^2 \equiv 0 \pmod 4 \Rightarrow 14y^2 \equiv 0 \pmod 8$.
$15x^2-71z^2 \equiv 7x^2-7z^2 \equiv (x-z)(x+z) \pmod 8$.
If $x \equiv z \pmod 4$, then $x-z$ has a factor 4, and also x and z are both odd or both even, so $x+z$ has a factor 2, and $(x-z)(x+z) \equiv 0 \pmod 8$.

77 $3 \cdot 5 \cdot 7 \cdot 71 \cdot 4 \cdot 2 = 4 \cdot 15 \cdot 14 \cdot 71$.

78 Analogue of q 44.

80 If $|15x_1^2+14y_1^2-71z_1^2| < k$ and $15x_1^2+14y_1^2-71z_1^2 \equiv 0 \pmod k$, then $15x_1^2+14y_1^2-71z_1^2=0$.

81 Since gcd $(x, y, z)=1$, all three may not be even. If two are even, three are even. So at most one is even. If x, y odd, then $3x^2$ and $5y^2$ are both odd so $3x^2+5y^2$ is even and z must be even. All three may not be odd. Exactly one is even.

82 Suppose x is even, then $5y^2=62z^2-3x^2$ is even so y is even, and then $3x^2+5y^2$ has a factor 4, so z is even which contradicts gcd $(x, y, z)=1$. Similarly for y, so x and y are both odd. $x=2n+1$, $x^2=4n^2+4n+1=4n(n+1)+1$ and since either n or $n+1$ is even, $x^2 \equiv 1 \pmod 8$. Thus

$3x^2+5y^2 \equiv 0 \pmod 8$, so $62z^2 \equiv 0 \pmod 8$ and $31z^2 \equiv 0 \pmod 4$ so z is even.
For any two odd numbers, x, y, $x \equiv \pm y \pmod 4$.

83 As in q 82, both x and y are odd, so $x^2 \equiv y^2 \equiv 1 \pmod 8$.
Now $2z^2 \equiv 3+7 \pmod 8$, so $z^2 \equiv 1 \pmod 8$ and z is odd.
As in q 82, $x \equiv \pm y \pmod 4$.

84 (i) Since $\gcd(ax_1, by_1, cz_1, 4abc) = 1$ at most one of ax_1 and by_1 can be even. But if ax_1 is even, then so is ax_1^2 and $by_1^2 \equiv -cz_1^2 - ax_1^2 \pmod 8$ so by_1^2 is also even, and by_1 is too.
(ii) If ax_1 is odd, then so is x_1, so $x_1^2 \equiv y_1^2 \equiv 1 \pmod 8$. Therefore $ax_1^2 + by_1^2 + cz_1^2 \equiv a+b+cz_1^2 \equiv 0 \pmod 8$.
So if z_1 is even, $a+b \equiv 0 \pmod 8$.
(iii) If z_1 is odd, $z_1^2 \equiv 1 \pmod 8$, so $a+b+c \equiv 0 \pmod 8$.
If $x \equiv y \equiv 0 \pmod 4$, then z is even and each of ax^2, by^2 and $cz^2 \equiv 0 \pmod 8$.
If $x \equiv y \equiv 2 \pmod 4$ and z_1 is even, $ax^2+by^2+cz^2 \equiv 4a+4b \equiv 0 \pmod 8$.
If $x \equiv y \equiv 2 \pmod 4$ and z_1 is odd, $ax^2+by^2+cz^2 \equiv 4a+4b+4c \equiv 0 \pmod 8$.
If $x \equiv y \equiv \pm 1 \pmod 4$ and z_1 is even, $ax^2+by^2+cz^2 \equiv a+b \equiv 0 \pmod 8$.
If $x \equiv y \equiv \pm 1 \pmod 4$ and z_1 is odd, $ax^2+by^2+cz^2 \equiv a+b+c \equiv 0 \pmod 8$.

85 Since $p|a$, $by_1^2+cz_1^2 \equiv 0 \pmod p$. If one of these terms had a factor p, both would have, but this would conflict with $\gcd(ax_1, by_1, cz_1, 4abc) = 1$. Since z_1 does not have a factor p, there exists an integer k such that $kz_1 \equiv 1 \pmod p$. Now take $l = y_1 k$.

86 The lattice is formed by the intersection of one lattice for each prime factor of $4abc$.
For each odd prime factor p of a or b or c, we construct a lattice $y \equiv lz \pmod p$ or $z \equiv lx$ or $x \equiv ly \pmod p$ as in n 85. Exactly one of ax_1, by_1 and cz_1 is even, so at most one of a, b and c is even. Suppose c is even, then q 84 shows how to construct the necessary lattice with index 8. If a, b, c are all odd, then z_1 is even, so $a+b \equiv 0 \pmod 4$ and we may take a lattice of index 4 with $x \equiv y \pmod 2$ and $z \equiv 0 \pmod 2$. The ellipsoid has volume $\frac{4}{3}\pi \cdot 8|abc| > 8 \cdot 4|abc|$, and by q 73 contains a lattice point other than the origin. For this point we have $ax^2+by^2+cz^2 \equiv 0 \pmod{4abc}$ and $|ax^2+by^2+cz^2| < |4abc|$, so $ax^2+by^2+cz^2 = 0$.
The statement proved here is not the most usual form of Legendre's theorem. Three forms are given by Nagell (1964). A simple proof that if a solution to $ax^2+by^2+cz^2=0$ exists there is a solution in which $|x| < \sqrt{|bc|}$, $|y| < \sqrt{|ca|}$ and $|z| < \sqrt{|ab|}$, was given by L. J. Mordell in *Journal of Number Theory* (1969), vol. 1.

Historical note

The first use of lattices to obtain a number theoretical result is the contribution that F. M. G. Eisenstein made in 1844 to a proof of the law of quadratic reciprocity and which we have used in chapter 4. But it was H. Minkowski who developed the geometry into the powerful tool which it has now become. His theorem, the simplest forms of which we have explored, was first published in 1891.

In 1785, A. M. Legendre proved that if no two of a, b and c have a common factor, and if each is neither zero nor divisible by a square, then $ax^2+by^2+cz^2=0$ has integral solutions, not all zero, if and only if $-bc$, $-ac$ and $-ab$ are quadratic residues modulo a, b and c respectively and a, b, and c are not all of the same sign. The application of the geometry of numbers to quadratic forms was touched on by Eisenstein and developed by Minkowski.

10

Continued fractions

A pocket calculator is needed for this and the following chapter.

Irrational square roots

1 If a and b are positive real numbers and $a/b=\sqrt{2}$, prove that $(2b-a)/(a-b)=\sqrt{2}$ and that $b>a-b>0$.
Use Fermat's method of descent to prove that a and b cannot both be integers.

2 If a and b are positive real numbers and $a/b=\sqrt{7}$, prove that $(7b-2a)/(a-2b)=\sqrt{7}$ and that $b>a-2b>0$. Deduce that a and b cannot both be integers.

3 Construct a proof, similar to the ones you have used in q 1 and q 2, which establishes that $\sqrt{57}$ is not a rational number.

4 Find integers a, b, c, d such that $2520/735=2^a 3^b 5^c 7^d$.

5 If $p_1, p_2, p_3, \ldots$ is the sequence of distinct prime numbers, explain why every non-zero rational number can be expressed in the form $\pm p_1^{a_1} p_2^{a_2} \ldots p_n^{a_n}$ for some n and uniquely defined integers $a_1, a_2, \ldots, a_n$.

6 If $r=\pm p_1^{a_1} p_2^{a_2} \ldots p_n^{a_n}$, what can be said about the indices for r^2?

7 Can 2, 3, 5 or 6 be the square of a rational number?

Convergence

The notion of a continued fraction emerges from an attempt to find rational approximations to irrational square roots.

8 Prove that $\sqrt{2}=1+\dfrac{1}{1+\sqrt{2}}$, and deduce that

$$\sqrt{2}=1+\cfrac{1}{2+\cfrac{1}{1+\sqrt{2}}}=1+\cfrac{1}{2+\cfrac{1}{2+\cfrac{1}{1+\sqrt{2}}}}.$$

9 The five numbers

$$1,\quad 1+\tfrac{1}{2},\quad 1+\cfrac{1}{2+\frac{1}{2}},\quad 1+\cfrac{1}{2+\cfrac{1}{2+\frac{1}{2}}},\quad 1+\cfrac{1}{2+\cfrac{1}{2+\cfrac{1}{2+\frac{1}{2}}}}$$

are also conventionally exhibited in the form

$$1,\quad 1+\tfrac{1}{2},\quad 1+\frac{1}{2+}\,\frac{1}{2},\quad 1+\frac{1}{2+}\,\frac{1}{2+}\,\frac{1}{2},\quad 1+\frac{1}{2+}\,\frac{1}{2+}\,\frac{1}{2+}\,\frac{1}{2}$$

and in the form

[1], [1, 2], [1, 2, 2], [1, 2, 2, 2], [1, 2, 2, 2, 2].

(i) Express each of these five numbers as a quotient of integers.
(ii) If a/b and c/d are adjacent terms in (i), find $ad-bc$ for the four possible cases.
(iii) Use a calculator to express these five numbers in decimal form, and notice whether the sequence increases, decreases or oscillates, and whether it appears to approach $\sqrt{2}$.

10 For real numbers $a_1, a_2, a_3, \ldots, a_n$, the simple continued fraction

$$a_1+\cfrac{1}{a_2+\cfrac{1}{a_3+\cfrac{\ddots}{\cfrac{1}{a_{n-1}+\cfrac{1}{a_n}}}}}$$

is denoted by $[a_1, a_2, a_3, \ldots, a_n]$. Convention demands that $a_i \geqslant 1$ when $i \geqslant 2$.

Evaluate each of the following as a quotient.
$[1, x]$, $[1, 2]$, $[1, 2+1/y]$, $[1, 2, y]$, $[1, 2, 3]$, $[1, 2, 3+1/z]$, $[1, 2, 3, z]$, $[1, 2, 3, 4]$, $[1, 2, 3, 4+1/u]$, $[1, 2, 3, 4, u]$.

If a/b and c/d are adjacent terms in the sequence [1], [1, 2], [1, 2, 3], [1, 2, 3, 4], [1, 2, 3, 4, 5] evaluate $ad-bc$ in the four possible cases.

⟦218⟧

11 If $[1, 2, 3, \ldots, 9, 10] = p/q$ and $[1, 2, 3, \ldots, 9, 10, 11] = r/s$, use q 10 to conjecture a value for $[1, 2, 3, \ldots, 9, 10, 11, x]$.

12 Let $p_1 = 1$, $p_2 = 3$ and $p_n = np_{n-1} + p_{n-2}$ for $n \geqslant 3$.
Let $q_1 = 1$, $q_2 = 2$ and $q_n = nq_{n-1} + q_{n-2}$ for $n \geqslant 3$. Verify that

$$[1, 2, x] = \frac{xp_2 + p_1}{xq_2 + q_1}, [1, 2, 3, x] = \frac{xp_3 + p_2}{xq_3 + q_2}$$

and prove by induction that

$$[1, 2, 3, \ldots, n-1, x] = \frac{xp_{n-1} + p_{n-2}}{xq_{n-1} + q_{n-2}}.$$

13 If $p_1 = a_1$, $p_2 = a_1a_2 + 1$, $p_n = a_np_{n-1} + p_{n-2}$ for $n \geqslant 3$, and $q_1 = 1$, $q_2 = a_2$, $q_n = a_nq_{n-1} + q_{n-2}$ for $n \geqslant 3$, verify that

$$[a_1, a_2, x] = \frac{xp_2 + p_1}{xq_2 + q_1}, \qquad [a_1, a_2, a_3, x] = \frac{xp_3 + p_2}{xq_3 + q_2}.$$

Prove by induction that

$$[a_1, a_2, \ldots, a_{n-1}, x] = \frac{xp_{n-1} + p_{n-2}}{xq_{n-1} + q_{n-2}}$$

and deduce that $[a_1, a_2, \ldots, a_n] = p_n/q_n$.

14 With the notation of q 13, must p_n/q_n be a rational number if $a_1, a_2, \ldots, a_n$ are positive integers?

15 With the notation of q 13

(i) show that $p_nq_{n-1} - p_{n-1}q_n = -(p_{n-1}q_{n-2} - p_{n-2}q_{n-1})$;
(ii) find a value for $p_2q_1 - p_1q_2$;
(iii) deduce that $p_nq_{n-1} - p_{n-1}q_n = (-1)^n$.

16 If a, m, n, u and v are positive real numbers and $m/n < u/v$, prove that

$$\frac{m}{n} < \frac{am+u}{an+v} < \frac{u}{v} \text{ and that}$$

$$\frac{m}{n} < \frac{m+au}{n+av} < \frac{u}{v}.$$

17 With the notation of q 13, must each term of the sequence

$$\frac{p_1}{q_1}, \frac{p_2}{q_2}, \ldots, \frac{p_n}{q_n}$$

lie between its two predecessors? Use the fact that

$$\frac{p_n}{q_n} - \frac{p_{n-1}}{q_{n-1}} = \frac{(-1)^n}{q_{n-1}q_n}$$

⟦218⟧

to prove that each even term of the sequence is greater than its predecessor and each odd term of the sequence less than its predecessor.

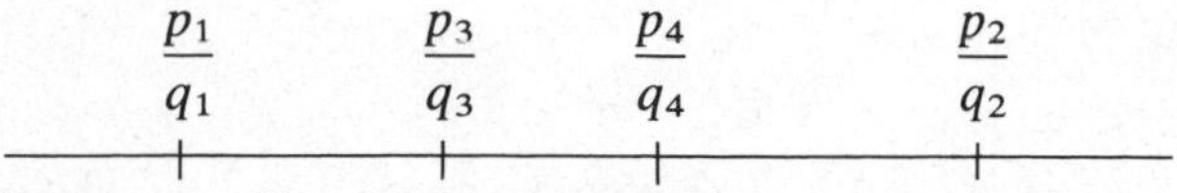

Is every odd term less than every even term?

18 Starting from the rational number $\frac{61}{48}$, use a calculator to verify that $\frac{61}{48}=[1, 3, 1, 2, 4]$.
The numbers $[1]$, $[1, 3]$, $[1, 3, 1]$, $[1, 3, 1, 2]$, $[1, 3, 1, 2, 4]$ are called *convergents* to $\frac{61}{48}$.
Evaluate each of these convergents as a quotient of integers and use the penultimate convergent to find integers x and y such that $61x-48y=-1$.
Explain why $[1, 3, 1, 2, 4]=[1, 3, 1, 2, 3, 1]$.
Evaluate $[1, 3, 1, 2, 3]$ and find integers x and y such that $61x-48y=1$.
Since the x you have found is positive and less than 48 and the y you have found is positive and less than 61, prove that these are the least positive integers satisfying the equation.

19 Use the equations

$$168=2\cdot 73+22$$
$$73=3\cdot 22+7$$
$$22=3\cdot 7+1$$
$$7=7\cdot 1$$

to find integers a_1, a_2, a_3, a_4 such that
$\frac{168}{73}=[a_1, a_2, a_3, a_4]$.
Find $[a_1, a_2, a_3]$ and deduce an integral solution of $168x-73y=1$.
For what value of b does $\frac{168}{73}=[a_1, a_2, a_3, b, 1]$?
Evaluate $[a_1, a_2, a_3, b]$ and deduce an integral solution of $168x-73y=-1$.

20 Find positive integers $b_1, b_2, b_3, b_4, r_1, r_2, r_3$ with $25>r_1>r_2>r_3$ such that

$$217=2\cdot 96+25,$$
$$96=b_1\cdot 25+r_1,$$
$$25=b_2\cdot r_1+r_2,$$
$$r_1=b_3\cdot r_2+r_3,$$
$$r_2=b_4\cdot r_3,$$

⟦219⟧

and use these values to find integers a_1, a_2, a_3, a_4, a_5 such that $\frac{217}{96}=[a_1, a_2, a_3, a_4, a_5]$.

21 If a and b are positive integers with no common factor, must there exist positive integers $a_1, a_2, \ldots, a_n$ for some n such that

$a/b=[a_1, a_2, \ldots, a_n]$

where, alone among the a_i, a_1 can also be zero? How might you find the a_i, given a and b?

22 If $[a_1, a_2, \ldots, a_n]=[b_1, b_2, \ldots, b_m]$ and all the terms are positive integers except possibly for a_1 and b_1 which may be any integers, explain why $[a_2, \ldots, a_n]$ and $[b_2, \ldots, b_m]$ are both greater than 1 if $m, n \geqslant 2$, so that

$$\frac{1}{[a_2, \ldots, a_n]} \quad \text{and} \quad \frac{1}{[b_2, \ldots, b_m]}$$

are both less than 1.
Deduce that a_1 and b_1 are the integral parts of the left and right hand sides of the initial equation, so that $a_1=b_1$ and $[a_2, \ldots, a_n]=[b_2, \ldots, b_m]$. If this argument is repeated n times, and $n \leqslant m$, we have $a_i=b_i$ for $i<n$, and $[a_n]=[b_n, \ldots, b_m]$. This is an equation of integers so either $n=m$ and $a_n=b_n$, or $n+1=m$ and $a_n=b_n+1$, $b_m=b_{n+1}=1$.

23 Let $a_1=[\sqrt{3}]$ as in q 4.44 and define x_2 by $\sqrt{3}=[a_1, x_2]$.
Let $a_2=[x_2]$ as in q 4.44 and define x_3 by $\sqrt{3}=[a_1, a_2, x_3]$.
Let $a_3=[x_3]$ as in q 4.44 and define x_4 by $\sqrt{3}=[a_1, a_2, a_3, x_4]$.
In general let

$a_{n-1}=[x_{n-1}]$ and define x_n by $\sqrt{3}=[a_1, a_2, a_3, \ldots, a_{n-1}, x_n]$.

Use the fact that

$$\sqrt{3}=1+\cfrac{1}{1+\cfrac{1}{1+\sqrt{3}}}$$

to determine a_1, a_2, a_3, a_4, a_5, a_6 and a_7. The rational numbers $[a_1]$, $[a_1, a_2]$, $[a_1, a_2, a_3]$, $[a_1, a_2, a_3, a_4]$, $\ldots$, $[a_1, a_2, \ldots, a_n]$ are called the first n *convergents* to $\sqrt{3}$.
Find the first five convergents to $\sqrt{3}$ and illustrate them in order on a number line.
Must $\sqrt{3}$ have an infinity of convergents?

24 For any irrational number x, let $a_1=[x]$ as in q 4.44, and define x_2 by $x=[a_1, x_2]$. Let $a_2=[x_2]$ as in q 4.44, and define x_3 by $x= [a_1, a_2, x_3]$.

By recursion let

$a_{n-1}=[x_{n-1}]$ and define x_n by $x=[a_1, a_2, \ldots, a_{n-1}, x_n]$.

Then $[a_1]$, $[a_1, a_2]$, $[a_1, a_2, a_3], \ldots, [a_1, a_2, \ldots, a_n], \ldots$ are called the convergents to x.

Explain why $0<1/x_2<1$ and deduce that $x_2>1$.

Explain why $0<1/x_3<1$ and deduce that $x_3>1$.

Explain why $a_2, a_3, \ldots$ are positive integers.

Writing $c_n=[a_1, a_2, \ldots, a_n]$ deduce from q 17 that

$c_1<c_3<c_5<c_6<c_4<c_2$.

Using the notation of q 13, explain why

$$x=\frac{x_n p_{n-1}+p_{n-2}}{x_n q_{n-1}+q_{n-2}}$$

and deduce that x lies between any two of its successive convergents

$c_1<c_3<c_5<x<c_6<c_4<c_2$.

With the help of q 17 show that

$$\left|x-\frac{p_n}{q_n}\right|<\frac{1}{q_n q_{n+1}}.$$

Since q_n is an increasing and unbounded sequence of positive integers, this proves that $p_n/q_n \to x$ as $n \to \infty$.

25 From any infinite continued fraction $[a_1, a_2, \ldots, a_n, \ldots]$ where the a_i are positive integers when $i>1$ and a_1 is an integer, it is possible to construct a sequence of convergents

$$\begin{aligned}
c_1 &= [a_1]=a_1,\\
c_2 &= [a_1, a_2],\\
c_3 &= [a_1, a_2, a_3],\\
&\vdots\\
c_n &= [a_1, a_2, \ldots, a_n],\\
&\vdots
\end{aligned}$$

Use q 17 to prove that (c_{2n+1}), the subsequence of odd convergents, is monotonic increasing and bounded above, and to prove that (c_{2n}), the subsequence of even convergents, is monotonic decreasing and bounded below, so that each of these sequences has a limit. Use q 17 to prove that these limits may not be different. If $(c_n)\to\alpha$ as $n\to\infty$, use the fact that the integral part of $c_n=a_1$ for all n, to show that $[\alpha]=a_1$. Use similar arguments to show that $[a_1, a_2, \ldots, a_n, \ldots]$ is the continued fraction for α as defined in q 24.

⟦221⟧

Purely periodic continued fractions

Now that we know that each simple continued fraction converges to a unique real number, we start to enquire as to which real numbers are the limits of those continued fractions with a purely periodic sequence of entries.

26 If $\alpha = [1, 1, 1, \ldots]$, use the fact that $\alpha = [1, \alpha]$ to show that $\alpha^2 - \alpha - 1 = 0$, and hence find the value of α.

27 If $\alpha = [2, 2, 2, \ldots]$, use the fact that $\alpha = [2, \alpha]$ to show that $\alpha^2 - 2\alpha - 1 = 0$.

28 If $\alpha = [a, a, a, \ldots]$, where a is a positive integer, use the fact that $\alpha = [a, \alpha]$ to find a quadratic equation which is satisfied by α, and hence a formula for α.

29 If $\alpha = [1, 2, 1, 2, 1, 2, \ldots]$, use the fact that $\alpha = [1, 2, \alpha]$ to prove that $2\alpha^2 - 2\alpha - 1 = 0$ and hence find the value of α.

30 If $\beta = [2, 1, 2, 1, 2, 1, \ldots]$, or briefly $\beta = [\dot{2}, \dot{1}]$, use the fact that $\beta = [2, 1, \beta]$ to prove that $\beta^2 - 2\beta - 2 = 0$, and hence find the value of β. Show that the roots of $2x^2 - 2x - 1 = 0$ are $[\dot{1}, \dot{2}]$ and $-1/[\dot{2}, \dot{1}]$.

31 If $\alpha = [a, b, a, b, \ldots]$, or briefly $[\dot{a}, \dot{b}]$, where a and b are positive integers, find a quadratic equation which is satisfied by α. Explain why the quadratic equation which you have found has one positive and one negative root. Find a quadratic equation which is satisfied by $[b, a, b, a, \ldots] = [\dot{b}, \dot{a}]$ and show that $[\dot{a}, \dot{b}]$ and $-1/[\dot{b}, \dot{a}]$ satisfy the same quadratic equation with integral coefficients.

32 If $\alpha = [a, b, c, a, b, c, \ldots]$, or briefly $\alpha = [\dot{a}, b, \dot{c}]$, where a, b and c are positive integers, find a quadratic equation with integral coefficients which is satisfied by α. Write down a quadratic equation which is satisfied by $[c, b, a, c, b, a \ldots]$ and deduce that $[\dot{a}, b, \dot{c}]$ and $-1/[\dot{c}, b, \dot{a}]$ are roots of the same quadratic equation with integral coefficients.

33 If $\alpha = [a_1, a_2, \ldots, a_n, \alpha]$, where $a_1, a_2, \ldots, a_n$ are positive integers, $[a_1, a_2, \ldots, a_{n-1}] = p_{n-1}/q_{n-1}$ and $[a_1, a_2, \ldots, a_n] = p_n/q_n$, use q 13 to prove that α satisfies a quadratic equation with integral coefficients.
Deduce that any purely periodic continued fraction, for example $[\dot{a}_1, a_2, a_3, \ldots, \dot{a}_n]$, where the entries $a_1, a_2, \ldots, a_n$ are all positive integers, satisfies a quadratic equation with integral coefficients.

[[221]]

34 With the notation of q 13, show that

$p_2/p_1=[a_2, a_1]$ and that $p_3/p_2=[a_3, a_2, a_1]$.

Use the equation $p_n=a_np_{n-1}+p_{n-2}$, for $n\geqslant 3$, to establish by induction that

$p_n/p_{n-1}=[a_n, a_{n-1}, \ldots, a_2, a_1]$ when $n\geqslant 2$.

35 With the notation of q 13, show that

$q_2/q_1=a_2$ and that $q_3/q_2=[a_3, a_2]$.

Use the equation $q_n=a_nq_{n-1}+q_{n-2}$, for $n\geqslant 3$, to establish by induction that

$q_n/q_{n-1}=[a_n, a_{n-1}, \ldots, a_2]$ for $n\geqslant 2$.

36 If $\alpha=[\dot{a}_1, \dot{a}_2, \ldots, \dot{a}_n]$ and $\beta=[\dot{a}_n, \dot{a}_{n-1}, \ldots, \dot{a}_1]$, where the a_i are positive integers, and if

$p_i/q_i=[a_1, a_2, \ldots, a_i]$ for $i\leqslant n$,

prove that

$$\alpha=\frac{\alpha p_n+p_{n-1}}{\alpha q_n+q_{n-1}} \quad \text{and} \quad \beta=\frac{\beta p_n+q_n}{\beta p_{n-1}+q_{n-1}}.$$

Obtain quadratic equations with integral coefficients satisfied by α and β.

37 If $\alpha=[\dot{a}_1, \dot{a}_2, \ldots, \dot{a}_n]$, deduce from q 36, that the roots of the quadratic equation derived from $\alpha=[a_1, a_2, \ldots, a_n, \alpha]$ are α and $-1/\beta$. Presuming that the a_i are positive integers, prove that one of these roots is greater than 1 and the other between -1 and 0.

38 Find a quadratic equation with integral coefficients for which $1+\sqrt{2}$ is a root. What is the other root of this equation? Is there more than one such equation? If so, what is the second root in each case?

39 The set of all numbers of the form $a+b\sqrt{2}$ where a and b are rational is denoted by $\mathbf{Q}(\sqrt{2})$. Is every number of this form, with $b\neq 0$, the root of a quadratic equation with integral coefficients?

40 If a and b are rational numbers and $a+b\sqrt{2}=0$, why must both a and $b=0$?
If a, b, c and d are rational numbers and $a+b\sqrt{2}=c+d\sqrt{2}$, why must $a=c$ and $b=d$?

41 Must the sum, difference, product and quotient of two numbers in $\mathbf{Q}(\sqrt{2})$ lie within the set (excepting, of course, division by zero)?

⟦222⟧

42 If a and b are rational numbers and $a+b\sqrt{2}$ is a root of the equation $px^2+qx+r=0$, where p, q, and r are integers, must $a-b\sqrt{2}$ also be a root of this equation?

43 If a and b are rational numbers and $\alpha=a+b\sqrt{2}$, we denote the *conjugate* of α, $a-b\sqrt{2}$, by $\bar{\alpha}$.
Is $\overline{\alpha+\beta}=\bar{\alpha}+\bar{\beta}$ for all $\alpha, \beta \in \mathbf{Q}(\sqrt{2})$?
Is $\overline{\alpha\beta}=\bar{\alpha}\bar{\beta}$ for all $\alpha, \beta \in \mathbf{Q}(\sqrt{2})$?
Is $\overline{1/\alpha}=1/\bar{\alpha}$ for all $\alpha \neq 0$ in $\mathbf{Q}(\sqrt{2})$?

44 Would the results you obtained in answering q 39–43 still hold if $\sqrt{2}$ were replaced by $\sqrt{d}$, where d is any non-square positive integer?

45 If $1+\sqrt{2}=\alpha$, find α_2 where $\alpha=[\alpha]+1/\alpha_2$.
Check whether $\alpha, \alpha_2>1$ and whether their conjugates are between -1 and 0.

46 If $2+\sqrt{7}=\beta$, find β_2 where $\beta=[\beta]+1/\beta_2$.
Check whether $\beta, \beta_2>1$ and whether their conjugates are between -1 and 0.
Find quadratic equations with integral coefficients which are satisfied by β and β_2 respectively.
Compare the discriminants of these two equations.

47 If α is an irrational number in $\mathbf{Q}(\sqrt{d})$ and $\alpha=[\alpha]+1/\alpha_2$, prove that α_2 is an irrational number in $\mathbf{Q}(\sqrt{d})$. If, in addition $1<\alpha$ and $-1<\bar{\alpha}<0$, prove that $1<\alpha_2$ and $-1<\bar{\alpha}_2<0$.

48 If the argument of q 38 is generalised, we can establish that every irrational in $\mathbf{Q}(\sqrt{d})$ is the root of a quadratic equation with integral coefficients which is unique, except for trivial multiples. If $\alpha=n+1/\alpha_2$, where α is an irrational number in $\mathbf{Q}(\sqrt{d})$ and n is an integer, $a\alpha^2+b\alpha+c=0$ and $a'\alpha_2^2+b'\alpha_2+c'=0$, where a, b, c are integers with no common factor, and a', b', c' are integers with no common factor, prove that $b^2-4ac=b'^2-4a'c'$.

49 If $\alpha=(p+\sqrt{2})/q$, what integers p and q ensure that $1<\alpha$ and $-1<\bar{\alpha}<0$?

50 If $\alpha=(p+\sqrt{11})/q$, what integers p and q ensure that $1<\alpha$ and $-1<\bar{\alpha}<0$?

51 If $\alpha=(p+\sqrt{d})/q$, where p and q are integers and $\alpha>1$, $0>\bar{\alpha}>-1$, show that $\sqrt{d}>p>0$, and $2\sqrt{d}>q>0$, so that the conditions constrain p and q to a finite number of integral values.

52 The name *quadratic irrational* is given to those real numbers which are solutions of a quadratic equation with integral coefficients, but which are not rational.
We have already proved in q 37 that every purely periodic continued fraction converges to a quadratic irrational α such that $\alpha > 1$, $0 > \bar{\alpha} > -1$, and we shall next establish the converse.
If α is a quadratic irrational such that $\alpha > 1$, $0 > \bar{\alpha} > -1$, show that there exist positive integers p, q and d, with d a non-square such that $\alpha = (p + \sqrt{d})/q$.

53 If d is a non square positive integer, p and q are integers, $\alpha = (p + \sqrt{d})/q$, $\alpha > 1$ and $0 > \bar{\alpha} > -1$, show that
(i) if $\alpha = [a_1, \alpha_2]$, where $a_1 = [\alpha]$, then $\alpha_2 = (p_1 + \sqrt{d})/q_1$, where p_1 and q_1 are integers, using q 48;
(ii) if $\alpha = [a_1, a_2, \alpha_3]$, where $a_2 = [\alpha_2]$, then $\alpha_3 = (p_2 + \sqrt{d})/q_2$, where p_2 and q_2 are integers;
(iii) if $\alpha = [a_1, a_2, \ldots, a_{n-1}, \alpha_n]$, where $a_{n-1} = [\alpha_{n-1}]$, then $\alpha_n = (p_{n-1} + \sqrt{d})/q_{n-1}$ for some integers p_{n-1} and q_{n-1};
(iv) by q 47, each $\alpha_i > 1$ and $0 > \bar{\alpha}_i > -1$;
(v) by q 51, there are only a finite number of distinct α_i;
(vi) if $\alpha_n = \alpha_m$ is the first repetition, $1 \leq n < m$, then $\alpha_{n+1} = \alpha_{m+1}$, and by induction $\alpha_{n+i} = \alpha_{m+i}$ for all positive integers i;
(vii) if $\beta_n = -1/\bar{\alpha}_n$, then $\beta_n > 1$ for all n;
(viii) if $n > 1$, then $\beta_n = a_{n-1} + 1/\beta_{n-1}$, so $a_{n-1} = [\beta_n]$;
(ix) $\beta_n = \beta_m$, so $[\beta_n] = [\beta_m]$, and $\beta_{n-1} = \beta_{m-1}$, so $\alpha_{n-1} = \alpha_{m-1}$, and consequently $n = 1$.

Pell's equation

Now that we have an exact knowledge of the relationship between purely periodic continued fractions and quadratic irrationals we can use this knowledge to determine those quadratic irrationals which are the limits of continued fractions periodic after the first term. We first determine the continued fractions for the square roots of positive integers and apply the results in solving Pell's equation, $x^2 - dy^2 = 1$.

54 For what value of the integer n is $n + \sqrt{2} > 1$ *and* $0 > n - \sqrt{2} > -1$?
Give the periodic continued fraction of $n + \sqrt{2}$ for this value of n, and deduce the continued fraction for $\sqrt{2}$.

55 Repeat the procedure of q 54 to find continued fractions for $\sqrt{3}$, $\sqrt{5}$, $\sqrt{6}$ and $\sqrt{7}$.

⟦224⟧

56 If d is a non-square positive integer, must $[\sqrt{d}]+\sqrt{d}$ have an expansion as a purely periodic continued fraction?

57 Use q 56 to say why the continued fraction for $\sqrt{d}$ is periodic after the first term when d is a non-square positive integer.

58 Why must $5+\sqrt{33}$ give a purely periodic continued fraction?
If $5+\sqrt{33}=[\dot{a}_1, a_2, \ldots, \dot{a}_n]$, where the a_i are positive integers, what is a_1?
Deduce that $\sqrt{33}=[5, \dot{a}_2, a_3, \ldots, a_n, 1\dot{0}]$,
so that

$$\sqrt{33}-5=[0, \dot{a}_2, a_3, \ldots, a_n, 1\dot{0}]$$

and

$$\frac{1}{\sqrt{33}-5}=[\dot{a}_2, a_3, \ldots, a_n, 1\dot{0}].$$

From q 37 and a generalised form of q 42, show that

$$\frac{1}{\sqrt{33}-5}=[\dot{a}_n, a_{n-1}, \ldots, a_2, \dot{a}_1].$$

Hence, $a_2=a_n$, $a_3=a_{n-1}$, etc.

59 If d is a non-square positive integer and $\sqrt{d}=[a_1, \dot{a}_2, a_3, \ldots, \dot{a}_n]$, determine the relation between a_1 and a_n, and which pairs within the period are equal.

60 Given that $\sqrt{14}=[3, \dot{1}, 2, 1, \dot{6}]$, prove that $\sqrt{14}=[3, 1, 2, 1, \sqrt{14}+3]$.
Given further that $[3, 1, 2]=\frac{11}{3}$ and $[3, 1, 2, 1]=\frac{15}{4}$, deduce that

$$\sqrt{14}=\frac{(\sqrt{14}+3)15+11}{(\sqrt{14}+3)4+3}.$$

Simplify this equation and check the truth of the result.

61 Given that d is a non-square positive integer and $\sqrt{d}=[a_1, \dot{a}_2, a_3, \ldots, a_n, 2\dot{a}_1]$, where the a_i are positive integers, prove that

$$\sqrt{d}=[a_1, a_2, a_3, \ldots, a_n, a_1+\sqrt{d}].$$

If $[a_1, a_2, \ldots, a_{n-1}]=p_{n-1}/q_{n-1}$, and $[a_1, a_2, \ldots, a_n]=p_n/q_n$ in the notation of q 13, show that

$$\sqrt{d}=\frac{(a_1+\sqrt{d})p_n+p_{n-1}}{(a_1+\sqrt{d})q_n+q_{n-1}}.$$

From the general form of q 40 deduce that $dq_n=a_1p_n+p_{n-1}$ and $a_1q_n+q_{n-1}=p_n$.
From q 15, deduce that $p_n^2-dq_n^2=(-1)^n$.

62 Use q 60 and q 61 to find integers x and y such that $x^2 - 14y^2 = 1$.

63 Use q 61 to find integers x and y such that

$$x^2 - 13y^2 = -1$$

given that $\sqrt{13} = [3, \dot{1}, 1, 1, 1, \dot{6}]$.
By working out convergents to $\sqrt{13}$ in its second period, find integers x and y such that

$$x^2 - 13y^2 = 1.$$

64 By considering the equation modulo 3, show that there are no integers x and y satisfying $x^2 - 3y^2 = -1$.
By considering the equation modulo 6, show that there are no integers x and y satisfying $x^2 - 6y^2 = -1$.

65 For any non-square positive integer d, must there exist non-zero integers x and y satisfying

$$x^2 - dy^2 = 1?$$

(*Pell's equation*)

66
$$\begin{aligned} 3^2 - 2 \cdot 2^2 = 1 &\Rightarrow (3 + 2\sqrt{2})(3 - 2\sqrt{2}) = 1 \\ &\Rightarrow (3 + 2\sqrt{2})^2(3 - 2\sqrt{2})^2 = 1^2 \\ &\Rightarrow (17 + 12\sqrt{2})(17 - 12\sqrt{2}) = 1 \\ &\Rightarrow 17^2 - 2 \cdot 12^2 = 1. \end{aligned}$$

Can this working be generalised, so that given $a^2 - 2b^2 = 1$, you can construct further integral solutions to $x^2 - 2y^2 = 1$?
Is there an infinity of pairs of integers x and y such that $x^2 - 2y^2 = 1$?

67 If $a + b\sqrt{2}$ is the smallest solution of $x^2 - 2y^2 = 1$ in integers (in the sense that if a, b, x, y are positive integers such that $a^2 - 2b^2 = x^2 - 2y^2 = 1$, then $1 < a + b\sqrt{2} \leqslant x + y\sqrt{2}$), since $1 < a + b\sqrt{2}$, $x + y\sqrt{2}$ must lie between two powers of $a + b\sqrt{2}$, say $(a + b\sqrt{2})^n \leqslant x + y\sqrt{2} < (a + b\sqrt{2})^{n+1}$. Multiply this inequality through by $(a - b\sqrt{2})^n$ to establish that $x + y\sqrt{2} = (a + b\sqrt{2})^n$.

68 If d is a non-square positive integer, must there be an infinity of integral solutions of

$$x^2 - dy^2 = 1,$$

and must all solutions be obtainable from the smallest solution?

69 If d is a non-square positive integer and a and b are the smallest positive integers such that $a^2 - db^2 = 1$, explain why

$$x = \tfrac{1}{2}[(a + b\sqrt{d})^n + (a - b\sqrt{d})^n]$$

and

$y=(1/2\sqrt{d})[(a+b\sqrt{d})^n-(a-b\sqrt{d})^n]$

are integers and why all solutions of $x^2-dy^2=1$ have this form.

70 Give a general formula for all integral solutions of $x^2-2y^2=1$ and deduce the form of all integral solutions of

$x^2+8xy+14y^2+6x+24y+8=0.$

71 Use your answer to q 55 to determine positive integers x and y such that $x^2-5y^2=1$. Deduce a solution to $x^2-5y^2=4$, in which both x and y are positive even integers. Find a solution to $x^2-5y^2=4$ in which both x and y are odd integers.
If $a^2-5b^2=4$ and $c^2-5d^2=4$, prove that $(ac+5bd)^2-5(ad+bc)^2=16$. Deduce that if

$$p+q\sqrt{5}=2\left(\frac{a+b\sqrt{5}}{2}\right)\left(\frac{c+d\sqrt{5}}{2}\right),$$

then $p^2-5q^2=4$. If a, b, c and d are positive integers, prove that p and q are integers.
Use this construction to obtain the solution you found before to $x^2-5y^2=4$ with both x and y even from the solution you found with x and y odd.

Lagrange's theorem on quadratic irrationals

Now we explore the irrational number corresponding to those continued fractions which are ultimately periodic. It was Lagrange who proved that these were just the quadratic irrationals.

72 If $\alpha=[1,\dot{2},\dot{3}]$, by reference to q 37, find an integer n such that $n+\alpha>1$ and $0>n+\bar{\alpha}>-1$.

73 Use q 53 to determine whether it is possible to find an integer n such that $n+\alpha>1$ and $0>n+\bar{\alpha}>-1$ when $\alpha=[1,2,\dot{3},\dot{4}]$.

74 Is the continued fraction expansion of

$\frac{1}{2}(3+\sqrt{12})$, $3+\sqrt{5}$, $3-\sqrt{5}$, $1+\frac{1}{4}\sqrt{3}$

(i) purely periodic,
(ii) periodic after the first term,
(iii) neither?

75 Use q 37 to give reasons why the continued fraction for $\frac{1}{4}\sqrt{3}$
(i) is not purely periodic
(ii) is not periodic after the first term.
If $\frac{1}{4}\sqrt{3}=[0,2,\alpha_3]$, explain why $\alpha_3>1$ and prove that $0>\bar{\alpha}_3>-1$. Deduce that the continued fraction for $\frac{1}{4}\sqrt{3}$ is periodic after the second term.

⟦225⟧

76 If α is a quadratic irrational and

$$\alpha=[a_1, a_2, \ldots, a_{n-1}, \alpha_n],$$

where the a_i are positive integers and α_n is a quadratic irrational, then

$$\alpha=\frac{\alpha_n p_{n-1}+p_{n-2}}{\alpha_n q_{n-1}+q_{n-2}}, \quad \text{with the conventions of q 13.}$$

Prove that

$$\bar{\alpha}_n=-\frac{q_{n-2}}{q_{n-1}}\frac{(p_{n-2}/q_{n-2}-\bar{\alpha})}{(p_{n-1}/q_{n-1}-\bar{\alpha})},$$

and with the help of q 24, deduce that $\bar{\alpha}_n$ is approximately $-q_{n-2}/q_{n-1}$ for sufficiently large n.
Deduce that for sufficiently large n, α_n is a periodic continued fraction and so α is ultimately periodic. (*Lagrange's theorem*)

77 In q 76 you found that the continued fraction for a quadratic irrational is ultimately periodic. Use q 33 and q 13 to prove the converse.

Automorphs of the indefinite form ax^2-by^2

Finally, in this chapter, we use continued fractions to explore the indefinite quadratic form ax^2-by^2 and to determine its group of automorphs.

78 $\sqrt{\frac{5}{3}}=[1, \dot{3}, \dot{2}]=\frac{1}{3}\sqrt{15}$. Find the first four convergents, $p_1/q_1, p_2/q_2, p_3/q_3, p_4/q_4$ to $\frac{1}{3}\sqrt{15}$. Let $\alpha=\frac{1}{3}\sqrt{15}=[1, \alpha_2]=[1, 3, \alpha_3]=[1, 3, 2, \alpha_4]$. Find integers $A_1, B_1, A_2, B_2, A_3, B_3, A_4, B_4$ such that

$$\alpha=\frac{A_1+\sqrt{15}}{B_1}, \qquad \alpha_2=\frac{A_2+\sqrt{15}}{B_2},$$

$$\alpha_3=\frac{A_3+\sqrt{15}}{B_3}, \qquad \alpha_4=\frac{A_4+\sqrt{15}}{B_4}.$$

Complete the table

n	1	2	3	4
$3p_n^2-5q_n^2$				
B_n				

79 Can any element of $\mathbf{Q}(\surd 15)$ be expressed in the form $(A+\surd N)/B$ for some choice of the integers A, B, N? Why can we in addition insist that $B|N-A^2$? When $B|N-A^2$, we shall refer to the quadratic irrational $(A+\surd N)/B$ as being in *standard form.*
If $\alpha=(A+\surd N)/B$ is in standard form, and $\alpha=n+1/\beta$ where n is an integer, show that there exist integers C and D that $\beta=(C+\surd N)/D$ in standard form. Deduce that if $\alpha_n=(A_n+\surd 15)/B_n$, extending the definitions of α_2, α_3, and α_4 in q 78, then α_n is in standard form.
Substitute this standard form in the equation

$$\alpha=\frac{\alpha_n p_{n-1}+p_{n-2}}{\alpha_n q_{n-1}+q_{n-2}}.$$

Use the general form of q 40, and q 15 to prove that
$3p_{n-1}^2-5q_{n-1}^2=(-1)^{n-1}B_n$.

80 Sketch the graph of $3x^2-5y^2=3$ for $|x|<5$, marking particularly those points with integral coordinates.
If $p^2-15k^2=1$, show algebraically that (a, b) lies on the graph if and only if $(pa+5kb,\ 3ka+pb)$ does.
Is

$$(x, y)\rightarrow(x, y)\begin{pmatrix} p & 3k \\ 5k & p \end{pmatrix}$$

a unimodular transformation when p and k are integers?
Evaluate the matrix product

$$\begin{pmatrix} p & 3k \\ 5k & p \end{pmatrix}\begin{pmatrix} 3 & 0 \\ 0 & -5 \end{pmatrix}\begin{pmatrix} p & 5k \\ 3k & p \end{pmatrix}$$

and describe the effect of the given transformation on the quadratic form $3x^2-5y^2$.
By considering the one–one correspondence

$$\begin{pmatrix} p & 3k \\ 5k & p \end{pmatrix}\rightarrow p+k\surd 15$$

and the argument of q 67, prove that the unimodular transformations of the form

$$(x, y)\rightarrow(x, y)\begin{pmatrix} p & 3k \\ 5k & p \end{pmatrix}$$

form an infinite cyclic group, a group of automorphs of the quadratic form $3x^2-5y^2$.

81 If a and b are positive integers, and $\gcd(a, b)=1$, prove that the quadratic form ax^2-by^2 has an infinity of automorphs.

82 If the unimodular transformation

$$(x, y) \to (x, y)\begin{pmatrix} p & q \\ r & s \end{pmatrix}$$

is an automorph of the quadratic form $3x^2 - 5y^2$, prove that
(i) $ps - qr = \pm 1$,
(ii) $3p^2 - 5q^2 = 3$,
(iii) $3r^2 - 5s^2 = -5$,
(iv) $3pr = 5qs$.
Use (i) and (iv) to prove that $(3p^2 - 5q^2)s = \pm 3p$, and deduce that $p^2 = s^2$.
Now use (ii) and (iii) to prove that $9r^2 = 25q^2$.
Deduce that the matrix of an automorph of $3x^2 - 5y^2$ has either the form $\begin{pmatrix} p & 3k \\ 5k & p \end{pmatrix}$ or the form $\begin{pmatrix} p & 3k \\ -5k & -p \end{pmatrix}$ where $p^2 - 15k^2 = 1$.

Notes and answers

For concurrent reading see bibliography: Olds (1963), Stark (1978), Chrystal (1964).

1 $\dfrac{2b-a}{a-b}=\dfrac{2-a/b}{a/b-1}=\dfrac{2-\sqrt{2}}{\sqrt{2}-1}=\sqrt{2}.$

Since $4>2>1$, $2>\sqrt{2}>1$ so $a>b$ and $a-b>0$. Also $2b>\sqrt{2}b=a$, so $b>a-b$.

The algebra establishes that if $\sqrt{2}$ could be expressed as a ratio of positive integers, then this can be done with a smaller denominator. Eventually we could have the denominator as 1 which would make $\sqrt{2}$ an integer.

2 $\dfrac{7b-2a}{a-2b}=\dfrac{7-2a/b}{a/b-2}=\dfrac{7-2\sqrt{7}}{\sqrt{7}-2}=\sqrt{7}$. Since $9>7>4$, $3>\sqrt{7}>2$.

Thus $a/b>2$ and $a-2b>0$. Also $3>a/b$, so $b>a-2b$.

3 Let $a/b=\sqrt{57}$. Since $64>57>49$, $8>\sqrt{57}>7$, so $b>a-7b>0$. Now $(57b-7a)/(a-7b)=\sqrt{57}$, and the argument used before may be applied.

4 $a=3$, $b=1$, $c=0$, $d=-1$.

5 If h and k are non-zero integers and the unique factorisation of these integers into primes is $h=\pm p_1^{b_1}p_2^{b_2}\dots p_m^{b_m}$ and $k=\pm p_1^{c_1}p_2^{c_2}\dots p_l^{c_l}$. Then the rational number $h/k=\pm p_1^{a_1}p_2^{a_2}\dots p_n^{a_n}$ where $n=\max\{m,l\}$ and $a_i=b_i-c_i$. If only one of the integers b_i or c_i exists, then $a_i=b_i$ or $a_i=-c_i$.

6 They are all even.

7 No, because of odd indices in prime factorisation.

9 (i) $\frac{1}{1}, \frac{3}{2}, \frac{7}{5}, \frac{17}{12}, \frac{41}{29}$.

(ii) $ad-bc=-1, 1, -1, 1$.

(iii) oscillates, approaches $\sqrt{2}$.

The word ‘simple’ in ‘simple continued fraction’ indicates the repeated numerator 1.

10 $\dfrac{x+1}{x}$, $\dfrac{3}{2}$, $\dfrac{3y+1}{2y+1}$, $\dfrac{3y+1}{2y+1}$, $\dfrac{10}{7}$, $\dfrac{10z+3}{7z+2}$,

$\dfrac{10z+3}{7z+2}$, $\dfrac{43}{30}$, $\dfrac{43u+10}{30u+7}$, $\dfrac{43u+10}{30u+7}$.

$ad-bc=-1, 1, -1, 1$.

11 $\dfrac{rx+p}{sx+q}$.

12 Suppose

$$[1,2,3,\ldots,n-1,x]=\frac{xp_{n-1}+p_{n-2}}{xq_{n-1}+q_{n-2}}.$$

Let $x=n+1/y$, then

$$[1,2,3,\ldots,n-1,n,y]=\frac{(n+1/y)p_{n-1}+p_{n-2}}{(n+1/y)q_{n-1}+q_{n-2}}$$

$$=\frac{y(np_{n-1}+p_{n-2})+p_{n-1}}{y(nq_{n-1}+q_{n-2})+q_{n-1}}$$

$$=\frac{yp_n+p_{n-1}}{yq_n+q_{n-1}}$$

This shows that a proof by induction is possible.

13 Proof as q 12. This result is the main tool for a practical and theoretical investigation of continued fractions, as it enables us to work them out from the front.

14 The definitions of p_n and q_n ensure that both are positive integers, so p_n/q_n is rational.

15 (i) $p_nq_{n-1}-p_{n-1}q_n=(np_{n-1}+p_{n-2})q_{n-1}-p_{n-1}(nq_{n-1}+q_{n-2})$.

(ii) 1.

(iii) (i) indicates that the absolute values are constant and the sign alternates.

16 $m/n<u/v\Rightarrow mv<nu\Rightarrow mv+amn<nu+amn$

$$\Rightarrow m(an+v)<n(am+u).$$

The result also follows by considering the parallelogram with vertices $(0,0)$, (an, am), $(an+v, am+u)$ and (v,u).

17 $\dfrac{p_n}{q_n}=\dfrac{a_np_{n-1}+p_{n-2}}{a_nq_{n-1}+q_{n-2}}\Rightarrow\dfrac{p_n}{q_n}$ lies between $\dfrac{p_{n-1}}{q_{n-1}}$ and $\dfrac{p_{n-2}}{q_{n-2}}$ by q 16.

Dividing q 15 (iii) by q_nq_{n-1} gives the equation. All the q_i are positive integers, so when n is even $p_n/q_n>p_{n-1}/q_{n-1}$. Thus $p_{2n}/q_{2n}>p_{2n-1}/q_{2n-1}$ and p_{2n+1}/q_{2n+1} lies between these two.

Moreover p_{2n+2}/q_{2n+2} lies between p_{2n+1}/q_{2n+1} and p_{2n}/q_{2n}, so $[p_{2n+1}/q_{2n+1}, p_{2n+2}/q_{2n+2}]$ is nested within $[p_{2n-1}/q_{2n-1}, p_{2n}/q_{2n}]$ and every odd term is less than every even term.

18 $\frac{1}{1}, \frac{4}{3}, \frac{5}{4}, \frac{14}{11}, \frac{61}{48}$.

$61\cdot 11-48\cdot 14=-1$.

$[1,3,1,2,3]=\frac{47}{37}$.

$61\cdot 37-48\cdot 47=1$.

If $61x-48y=61a-48b$, then $61(x-a)=48(y-b)$, and $x\equiv a \pmod{48}$, $y\equiv b \pmod{61}$.

19 $\frac{168}{73}=[2, 3, 3, 7]=[2, 3, 3, 6, 1]$.

$[2, 3, 3]=\frac{23}{10}$. $168 \cdot 10-73 \cdot 23=1$.

$[2, 3, 3, 6]=\frac{145}{63}$. $168 \cdot 63-73 \cdot 145=-1$.

20 $96=3 \cdot 25+21$,

$25=1 \cdot 21+4$,

$21=5 \cdot 4+1$,

$4=4 \cdot 1$.

$\frac{217}{96}=[2, 3, 1, 5, 4]$.

21 By the division algorithm (q 1.19),

$a=bq+r_1$ where $0 \leqslant r_1 < b$. Let $q=a_1$, then

$a/b=a_1+r_1/b$,

$$=a_1+\frac{1}{b/r_1}, \quad \text{if } r_1 \neq 0.$$

By the division algorithm

$b=r_1q'+r_2$ where $0 \leqslant r_2 < r_1$.

Let $q'=a_2$, then

$$\frac{b}{r_1}=a_2+\frac{r_2}{r_1}=a_2+\frac{1}{r_1/r_2}, \quad \text{if } r_2 \neq 0.$$

Again by the division algorithm $r_1=r_2q''+r_3$ where $0 \leqslant r_3 < r_2$ and we let $q''=a_3$.

Since the sequence $r_1, r_2, r_3, \ldots$ is a diminishing sequence of non-negative integers, $r_n=0$ for some n and then

$a/b=[a_1, a_2, \ldots, a_n]$.

Thus every positive rational number corresponds to a terminating simple continued fraction. By allowing a_1 alone to be negative it is possible to represent every rational number by a terminating continued fraction.

22 This establishes the uniqueness of the continued fraction expansion of a rational number except possibly in the last digit.

23 $\sqrt{3}=[1, 1, 2, 1, 2, 1, 2, \ldots]$.

$[1]=1, [1, 1]=2, [1, 1, 2]=\frac{5}{3}, [1, 1, 2, 1]=\frac{7}{4}, [1, 1, 2, 1, 2]=\frac{19}{11}$.

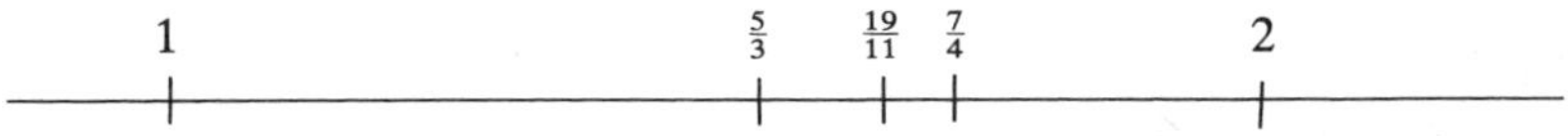

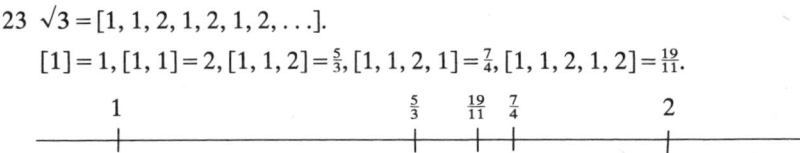

If the process of constructing convergents were to terminate, then $\sqrt{3}$ would be a rational number, and conversely.

24 $x = a_1 + 1/x_2$, but $[x] = a_1$ and $0 < x - [x] < 1$ since x is irrational, so $0 < 1/x_2 < 1$.
$x_2 = a_2 + 1/x_3$, but $[x_2] = a_2$ and $0 < x_2 - [x_2] < 1$ since x_2 is irrational.
If $1 < x_n$, $1 \leqslant [x_n] = a_n$.
If $x = [a_1, a_2, \ldots, a_{n-1}, x_n]$, then by q 13,

$$x = \frac{x_n p_{n-1} + p_{n-2}}{x_n q_{n-1} + q_{n-2}}.$$

Since

$$x = [a_1, a_2, \ldots, a_n, a_{n+1}, x_{n+2}] = \frac{x_{n+2} p_{n+1} + p_n}{x_{n+2} q_{n+1} + q_n},$$

x lies between p_{n+1}/q_{n+1} and p_n/q_n, but

$$\left|\frac{p_{n+1}}{q_{n+1}} - \frac{p_n}{q_n}\right| = \frac{1}{q_n q_{n+1}}, \quad \text{so } \left|x - \frac{p_n}{q_n}\right| < \frac{1}{q_n q_{n+1}}.$$

25 Obviously $c_{2n-1} < c_{2n}$, but c_{2n+1} lies between these two so $c_{2n-1} < c_{2n+1}$ and the odd convergents are monotonic increasing. Similarly the even convergents are monotonic decreasing. Each even convergent is an upper bound for the odd convergents and vice-versa, for if not $c_{2m} \leqslant c_{2n+1}$ for some m and n. If $m > n$, $c_{2m+1} < c_{2m} \leqslant c_{2n+1}$ contradicts the fact that the sequence of odd convergents is increasing. If $m \leqslant n$, $c_{2m} \leqslant c_{2n+1} < c_{2n+2}$ contradicts the fact that the sequence of even convergents is decreasing. Since $|c_{2n+1} - c_{2n}| = 1/q_{2n+1}q_{2n}$, the gap between the terms of the two sequences tends to 0.

$$c_3 < \alpha < c_2 \Rightarrow a_2 < \frac{1}{\alpha - a_1} < a_2 + \frac{1}{a_3}$$

$$\Rightarrow a_2 = \left[\frac{1}{\alpha - a_1}\right], \quad \text{and so on.}$$

These arguments establish a one–one correspondence between the set of irrational numbers and the set of all sequences of positive integers, with the first term any integer, and thus a well-defined meaning for an infinite simple continued fraction.

26 $\alpha = 1 + 1/\alpha$ so $\alpha^2 = \alpha + 1$. $\alpha = (1 + \sqrt{5})/2$, since $\alpha > 0$.

27 $\alpha = 2 + 1/\alpha$ so $\alpha^2 = 2\alpha + 1$. $\alpha = 1 + \sqrt{2}$, since $\alpha > 0$.

28 $\alpha = a + 1/\alpha$ so $\alpha^2 = a\alpha + 1$. $\alpha = [a + \sqrt{(a^2 + 4)}]/2$ since $\alpha > 0$.

29 $\alpha = 1 + \dfrac{1}{2 + 1/\alpha} = 1 + \dfrac{\alpha}{2\alpha + 1}$, so $2\alpha^2 + \alpha = 3\alpha + 1$ and $2\alpha^2 - 2\alpha - 1 = 0$ so $\alpha = \dfrac{1 + \sqrt{3}}{2}$.

30 $\beta = 2 + \dfrac{1}{1+1/\beta} = 2 + \dfrac{\beta}{\beta+1}$, so $\beta^2+\beta = 3\beta+2$ or $\beta^2-2\beta-2=0$.

$\beta = 1+\sqrt{3}$. $2(-1/\beta)^2 - 2(-1/\beta) - 1 = -(1/\beta^2)(\beta^2-2\beta-2) = 0$.

31 $\alpha = a + \dfrac{1}{b+1/\alpha} = a + \dfrac{\alpha}{b\alpha+1}$, so $b\alpha^2 - ab\alpha - a = 0$.

The product of the roots of this equation is $-a/b$, so one root, α, is positive and the other negative.
If $\beta = [\dot{b}, \dot{a}]$, $a\beta^2 - ab\beta - b = 0$, and β is the one positive root of this equation. However $a - ab(1/\beta) - b(1/\beta^2) = 0$, so $b(-1/\beta)^2 - ab(-1/\beta) - a = 0$ and it follows that α and $-1/\beta$ are the roots of $bx^2 - abx - a = 0$.

32 $$\alpha = a + \cfrac{1}{b + \cfrac{1}{c+1/\alpha}} = a + \cfrac{1}{b + \cfrac{\alpha}{c\alpha+1}} = a + \frac{c\alpha+1}{(bc+1)\alpha + b}$$

So $(bc+1)\alpha^2 + (b-a-c-abc)\alpha - ba - 1 = 0$.
If $\beta = [\dot{c}, \dot{b}, \dot{a}]$ $(ba+1)\beta^2 + (b-a-c-abc)\beta - bc - 1 = 0$. So

$$(bc+1)\left(-\frac{1}{\beta}\right)^2 + (b-a-c-abc)\left(-\frac{1}{\beta}\right) - ba - 1 = 0,$$

and both α and $-1/\beta$ satisfy the quadratic equation $(bc+1)x^2 + (b-a-c-abc)x - ba - 1 = 0$. α and $-1/\beta$ are distinct roots of this equation since one is positive and the other negative.

33 $\alpha = \dfrac{\alpha p_n + p_{n-1}}{\alpha q_n + q_{n-1}}$, so $q_n\alpha^2 + (q_{n-1} - p_n)\alpha - p_{n-1} = 0$.

34 $$\frac{p_2}{p_1} = \frac{a_1a_2+1}{a_1} = a_2 + \frac{1}{a_1} = [a_2, a_1],$$

$$\frac{p_3}{p_2} = \frac{a_3p_2+p_1}{p_2} = a_3 + \frac{1}{p_2/p_1} = [a_3, a_2, a_1].$$

By assuming $\dfrac{p_n}{p_{n-1}} = [a_n, a_{n-1}, \ldots, a_2, a_1]$,

$$\frac{p_{n+1}}{p_n} = \frac{a_{n+1}p_n + p_{n-1}}{p_n} = a_{n+1} + \frac{1}{p_n/p_{n-1}} = [a_{n+1}, a_n, \ldots, a_2, a_1].$$

35 $\dfrac{q_2}{q_1} = \dfrac{a_2}{1}$ and $\dfrac{q_3}{q_2} = \dfrac{a_3a_2+1}{a_2} = a_3 + \dfrac{1}{a_2} = [a_3, a_2]$.

If we assume that $q_n/q_{n-1} = [a_n, a_{n-1}, \ldots, a_2]$, then

$$\frac{q_{n+1}}{q_n} = \frac{a_{n+1}q_n + q_{n-1}}{q_n} = a_{n+1} + \frac{1}{q_n/q_{n-1}} = [a_{n+1}, a_n, \ldots, a_2].$$

36 The first n convergents to α are p_i/q_i, so

$$\alpha = \frac{\alpha p_n + p_{n-1}}{\alpha q_n + q_{n-1}}.$$

The $(n-1)$th and nth convergents to β are q_n/q_{n-1} and p_n/p_{n-1} respectively, so

$$\beta = \frac{\beta p_n + q_n}{\beta p_{n-1} + q_{n-1}}.$$

$$q_n\alpha^2 + (q_{n-1} - p_n)\alpha - p_{n-1} = 0.$$

$$p_{n-1}\beta^2 + (q_{n-1} - p_n)\beta - q_n = 0.$$

37 $q_n(-1/\beta)^2 + (q_{n-1} - p_n)(-1/\beta) - p_{n-1} = 0.$
Since $\alpha > a_1$ and $\beta > a_n$, $\alpha > 1$ and $-1 < -1/\beta < 0$.

38 If $x = 1 + \surd 2$, $(x-1)^2 = 2$, so $x^2 - 2x - 1 = 0$.
The other root is $1 - \surd 2$. If $1 + \surd 2$ is a root of a quadratic equation, this equation must be a multiple of $(x - 1 - \surd 2)(x - \alpha) = 0$. For integral coefficients, both $\alpha + 1 + \surd 2$ and $\alpha(1 + \surd 2)$ must be integers. If $\alpha + 1 + \surd 2$ is an integer a, $\alpha = a - 1 - \surd 2$ and $\alpha(1 + \surd 2) = a - 3 + (a-2)\surd 2$. Now $a = 2$ and $\alpha = 1 - \surd 2$.

39 If $x = m/n + (r/s)\surd 2$ where m, n, r and s are integers and r, n and $s \neq 0$, then $(x - m/n)^2 = 2r^2/s^2$ provides the required equation.

40 If $b \neq 0$, $\surd 2 = -a/b$ which would be rational. If $b = 0$, then $a = 0$ trivially.
$(a-c) + (b-d)\surd 2 = 0.$

41 Most of the results follow immediately from the closure of the rational numbers under these operations.

$$\frac{a + b\surd 2}{c + d\surd 2} = \frac{a + b\surd 2}{c + d\surd 2}\,\frac{c - d\surd 2}{c - d\surd 2} = \frac{ac - 2bd + (bc - ad)\surd 2}{c^2 - 2d^2}.$$

This number is in $\mathbf{Q}(\surd 2)$ unless both c and $d = 0$.

42 Let α and $a + b\surd 2$ be roots of $px^2 + qx + r = 0$, then
$p(x - \alpha)(x - a - b\surd 2) = px^2 + qx + r$, so

$$\alpha + a + b\surd 2 = -q/p, \quad \text{and} \quad \alpha(a + b\surd 2) = r/p.$$

From the first equation $\alpha = c - b\surd 2$ for some rational number c, and from the second equation $c = a$, using q 40.

43 $(a - b\surd 2) + (c - d\surd 2) = (a + c) - (b + d)\surd 2.$

$$(a - b\surd 2)(c - d\surd 2) = (ac + 2bd) - (ad + bc)\surd 2.$$

$$1/(a - b\surd 2) = \frac{a + b\surd 2}{a^2 - 2b^2} = \frac{a}{a^2 - 2b^2} + \frac{b}{a^2 - 2b^2}\surd 2.$$

$$1/(a + b\surd 2) = \frac{a - b\surd 2}{a^2 - 2b^2} = \frac{a}{a^2 - 2b^2} - \frac{b}{a^2 - 2b^2}\surd 2.$$

44 Yes, the only essential is that $\surd d$ be irrational.

45 $\alpha = \alpha_2 = 1 + \surd 2 > 1$. Since $-2 < -\surd 2 < -1$, $-1 < 1 - \surd 2 < 0$.

46 $\beta_2=\frac{1}{3}(2+\sqrt{7})$. Since $2<\sqrt{7}<3$, $-1<2-\sqrt{7}<0$ and $-1<\frac{1}{3}(2-\sqrt{7})<0$.
$(\beta-2)^2=7$, so $\beta^2-4\beta-3=0$; discriminant $4^2-4(-3)(1)=28$.
$(3\beta_2-2)^2=7$, so $9\beta_2^2-12\beta_2-3=0$ or $3\beta_2^2-4\beta_2-1=0$; discriminant $=$ 28.

47 If α_2 were rational, then since $[\alpha]$ is an integer, α would be rational. α_2 is in $\mathbf{Q}(\sqrt{d})$ by q 41 and q 44.
Since α is irrational, $0<\alpha-[\alpha]<1$, so $1<1/(\alpha-[\alpha])=\alpha_2$.

$$-1<\bar{\alpha}<0 \Rightarrow -1<[\alpha]+1/\bar{\alpha}_2<0, \quad \text{by q 43 and q 44,}$$
$$\Rightarrow -1-[\alpha]<1/\bar{\alpha}_2<-[\alpha],$$
$$\Rightarrow 0>\frac{-1}{1+[\alpha]}>\bar{\alpha}_2>-\frac{1}{[\alpha]}>-1, \quad \text{since } [\alpha]\geqslant 1.$$

48 Substituting $\alpha=n+1/\alpha_2$ in $a\alpha^2+b\alpha+c=0$ we obtain
$$(an^2+bn+c)\alpha_2^2+(2an+b)\alpha_2+a=0$$
and
$$(2an+b)^2-4a(an^2+bn+c)=b^2-4ac.$$
The procedure is like substituting $(nx+y, x)$ for (x, y) in the quadratic form $ax^2+bxy+cy^2$.

49 $p=1$, $q=1$ or 2.

50 $p=3$, $q=1, 2, 3, 4, 5, 6$;
$p=2$, $q=2, 3, 4, 5$;
$p=1$, $q=3, 4$.

51 Since $(p+\sqrt{d})/q>1>0>(p-\sqrt{d})/q$, q is positive. Since $|p+\sqrt{d}|>|p-\sqrt{d}|$, p is positive. Since $p+\sqrt{d}>0>p-\sqrt{d}$, $\sqrt{d}>-p>-\sqrt{d}$ and so $\sqrt{d}>p$. Since $p+\sqrt{d}>q$ and $\sqrt{d}>p$, $2\sqrt{d}>q$. Thus there are less than $2d$ possibilities for the pair (p, q).

52 If $a\alpha^2+b\alpha+c=0$, $\alpha=\dfrac{-b\pm\sqrt{(b^2-4ac)}}{2a}$.
We may presume that a is positive and that a, b and c have no common factor, without loss of generality. Then, since $\alpha>\bar{\alpha}$, we must take the positive root and, as in q 51, $-b$ is positive, so $\alpha=(p+\sqrt{d})/q$ where p and q are positive integers.

53 (i), (ii) and (iii) follow from q 48.
(vi) $a_n+1/\alpha_{n+1}=\alpha_n=\alpha_m=a_m+1/\alpha_{m+1}$, so $a_n=a_m$ and $\alpha_{n+1}=\alpha_{m+1}$.
(vii) $0>\bar{\alpha}_n>-1$ implies $-1/\bar{\alpha}_n>1$.
(viii) follows from q 43 and $\alpha_{n-1}=a_{n-1}+1/\alpha_n$.
Compare the result with q 33 and q 37.
We now know that every purely periodic continued fraction represents a quadratic irrational α such that $\alpha>1$ and $0>\bar{\alpha}>-1$.

54 $n=1$, $1+\sqrt{2}=[2,2,2,\ldots]$, $\sqrt{2}=[1,\dot{2}]$.

55 $1+\sqrt{3}=[2,1,\dot{2},\dot{1}]$, $\sqrt{3}=[1,\dot{1},\dot{2}]$.
$2+\sqrt{5}=[4,4,\dot{4}]$, $\sqrt{5}=[2,\dot{4}]$.
$2+\sqrt{6}=[4,2,\dot{4},\dot{2}]$, $\sqrt{6}=[2,\dot{2},\dot{4}]$.
$2+\sqrt{7}=[4,\dot{1},1,1,\dot{4}]$, $\sqrt{7}=[2,\dot{1},1,1,\dot{4}]$.

56 Yes, because if $\alpha=[\sqrt{d}]+\sqrt{d}$, $\alpha>1$, $0>\bar{\alpha}>-1$.

58 $0>5-\sqrt{33}>-1$. Since $6>\sqrt{33}>5$, $a_1=10$.
If $\alpha=5+\sqrt{33}$, $1/(\sqrt{33}-5)=-1/\bar{\alpha}$. The quadratic irrationals α and $\bar{\alpha}$ satisfy the same quadratic equation with integral coefficients.

59 $[\sqrt{d}]+\sqrt{d}$ is purely periodic, so $a_1=[\sqrt{d}]$, $a_n=2[\sqrt{d}]$.
$a_2=a_{n-1}$, $a_3=a_{n-2}$, etc.

60 Use q 13.

62 $15^2-14\cdot 4^2=1$.

63 Convergents are $\frac{3}{1}, \frac{4}{1}, \frac{7}{2}, \frac{11}{3}, \frac{18}{5}, \frac{119}{33}, \frac{137}{38}, \frac{256}{71}, \frac{393}{109}, \frac{649}{180}$. $18^2-13\cdot 5^2=-1$, $649^2-13\cdot 180^2=1$.

65 If the minimum period of $\sqrt{d}$ is even, then q 61 gives an affirmative answer. If the minimum period of $\sqrt{d}$ is odd, then q 61 again provides an affirmative answer taking $n=$ twice the length of the minimum period.

66
$$\begin{aligned} a^2-2b^2=1 &\Rightarrow (a+b\sqrt{2})(a-b\sqrt{2})=1 \\ &\Rightarrow (a+b\sqrt{2})^2(a-b\sqrt{2})^2=1^2 \\ &\Rightarrow (a^2+2b^2+2ab\sqrt{2})(a^2+2b^2-2ab\sqrt{2})=1 \\ &\Rightarrow (a^2+2b^2)^2-2(2ab)^2=1. \end{aligned}$$

Since $a^2+2b^2>a$, this solution is distinct from its predecessor, and the process may be repeated without possibility of repetition.

67 The product gives $1\leqslant(x+y\sqrt{2})(a-b\sqrt{2})^n<a+b\sqrt{2}$. But if $x^2-2y^2=1$ and $c^2-2d^2=1$, then $(x+y\sqrt{2})(c+d\sqrt{2})$ is also a solution. So $(x+y\sqrt{2})(a-b\sqrt{2})^n$ is a solution and by definition is equal to 1.

68 The argument of q 67 can be generalised.

69 If
$$\begin{aligned} x+y\sqrt{d}&=(a+b\sqrt{d})^n, \\ x-y\sqrt{d}&=(a-b\sqrt{d})^n \end{aligned}$$

by the general form of q 43. So all solutions have this form. Notice $x=\bar{x}$ and $y=\bar{y}$.

70 $x=\frac{1}{2}[(3+2\sqrt{2})^n+(3-2\sqrt{2})^n]$, $y=(1/2\sqrt{2})[(3+2\sqrt{2})^n-(3-2\sqrt{2})^n]$.
$(x+4y+3)^2-2y^2=1$.

71 $9^2-5\cdot4^2=1$, $18^2-5\cdot8^2=4$, $3^2-5\cdot1^2=4$.
$(a+b\sqrt{5})(a-b\sqrt{5})(c+d\sqrt{5})(c-d\sqrt{5})=16$,
so $[(a+b\sqrt{5})(c+d\sqrt{5})][(a-b\sqrt{5})(c-d\sqrt{5})]=16$,
and $2(p+q\sqrt{5})2(p-q\sqrt{5})=16$.
$a^2-5b^2=4\Rightarrow a\equiv b \pmod 2$. Now $ac+5bd\equiv 6ac\equiv 0 \pmod 2$.
Thus p is an integer, and from $p^2-5q^2=4$, q is also an integer.

$$2\left(\frac{3+\sqrt{5}}{2}\right)\left(\frac{3+\sqrt{5}}{2}\right)=7+3\sqrt{5} \text{ and } 2\left(\frac{3+\sqrt{5}}{2}\right)\left(\frac{7+3\sqrt{5}}{2}\right)=18+8\sqrt{5}.$$

72 $n=2$ makes $n+\alpha$ purely periodic.

73 If there were such an n, $n+\alpha$ would be purely periodic, so there is no such n.

74 $\frac{1}{2}(3+\sqrt{12})>3$, $0>\frac{1}{2}(3-\sqrt{12})>-1$; purely periodic.
$2+\sqrt{5}>4$, $0>2-\sqrt{5}>-1$; $3+\sqrt{5}$ is periodic after the first term.
$\overline{3-\sqrt{5}}>3-\sqrt{5}$; neither.
If $n+\frac{1}{4}\sqrt{3}>1$ then $n-\frac{1}{4}\sqrt{3}>0$: neither.

75 $\frac{1}{4}\sqrt{3}<1$, so not purely periodic.
$n+\frac{1}{4}\sqrt{3}>1\Rightarrow n\geqslant 1\Rightarrow n-\frac{1}{4}\sqrt{3}>0$, so not periodic after the first term.

76 Both p_{n-1}/q_{n-1} and p_{n-2}/q_{n-2} tend to α as $n\to\infty$ so $\bar{\alpha}_n$ tends to $-q_{n-2}/q_{n-1}$. But q_n is an unbounded increasing sequence of positive integers so $0>-q_{n-2}/q_{n-1}>1$. For $n>1$, $\alpha_n>1$ so ultimately $\alpha_n>1$, $0>\bar{\alpha}_n>-1$.

77 Every ultimately periodic continued fraction has the form $[a_1, a_2, \ldots, a_{n-1}, \alpha_n]$, where α_n is purely periodic. If $\alpha=[a_1, a_2, \ldots, a_{n-1}, \alpha_n]$, then

$$\alpha=\frac{\alpha_n p_{n-1}+p_{n-2}}{\alpha_n q_{n-1}+q_{n-2}} \text{ by q 13.}$$

But α_n is a quadratic irrational by q 33, so α lies in a $\mathbf{Q}(\sqrt{d})$ and since the continued fraction for α does not terminate, α is irrational by q 21.

78 $p_n/q_n=\frac{1}{1}, \frac{4}{3}, \frac{9}{7}, \frac{31}{24}$.
$\alpha_n=\frac{1}{3}(0+\sqrt{15}), \frac{1}{2}(3+\sqrt{15}), \frac{1}{3}(3+\sqrt{15}), \frac{1}{2}(3+\sqrt{15})$.
$3p_n^2-5q_n^2=-2, 3, -2, 3$.
$B_n=3, 2, 3, 2$.

79 $$\frac{a}{b}+\frac{c}{d}\sqrt{15}=\frac{ad\pm\sqrt{(15b^2c^2)}}{bd}; \frac{A+\sqrt{N}}{B}=\frac{AB\pm\sqrt{(NB^2)}}{B^2}; B^2|NB^2-(AB)^2;$$
$$\frac{A-\sqrt{N}}{B}=\frac{(-A)+\sqrt{N}}{(-B)}.$$

$$\beta=\frac{1}{\alpha-n}=\frac{B}{A+\sqrt{N}-nB}=\frac{B(-A+nB+\sqrt{N})}{N-(A-nB)^2}.$$

Now $B|N-A^2$, so $B|N-A^2+2nB-B^2$, and $D=[N-(A-nB)^2]/B$. Obviously $D|N-(A-nB)^2$, so with $C=-A+nB$, we have β in standard form. By equating rational parts,

$$5q_{n-1}=A_np_{n-1}+B_np_{n-2}.$$

By equating irrational parts,

$$3p_{n-1}=A_nq_{n-1}+B_nq_{n-2}.$$

The second equation times p_{n-1} minus the first equation times q_{n-1} gives the result.

80 $(\pm1, 0)$, $(\pm4, \pm3)$ lie on the curve, a hyperbola with asymptotes $y=\pm\sqrt{\tfrac{3}{5}}x$.

$(pa+5kb, 3ka+pb)$ lies on $3x^2-5y^2=3$,

if and only if

$$3(pa+5kb)^2-5(3ka+pb)^2=3,$$

or

$$(3p^2-45k^2)a^2-(5p^2-75k^2)b^2=3,$$

or

$$3a-5b^2=3.$$

The transformation is unimodular.

The matrix product is $\begin{pmatrix}3 & 0\\ 0 & -5\end{pmatrix}$, so the transformation leaves the quadratic form invariant, and is an automorph of the form. The correspondence is an isomorphism under the appropriate multiplications. All solutions of $p^2-15k^2=1$ are found from powers of the least solution; these form an infinite cyclic group under multiplication.

81 If $p^2-abk^2=1$,

$$\begin{pmatrix}p & ak\\ bk & p\end{pmatrix}\begin{pmatrix}a & 0\\ 0 & -b\end{pmatrix}\begin{pmatrix}p & bk\\ ak & p\end{pmatrix}=\begin{pmatrix}a & 0\\ 0 & -b\end{pmatrix},$$

and so $(x, y)\to(x, y)\begin{pmatrix}p & ak\\ bk & p\end{pmatrix}$ is an automorph of ax^2-by^2. The set of matrices of the form $\begin{pmatrix}p & ak\\ bk & p\end{pmatrix}$ under matrix multiplication is isomorphic to the numbers of the form $p+k\sqrt{ab}$ under multiplication, so any power of this matrix also provides an automorph.

82 Since $p^2=s^2$, $p=\pm s$. If $p=s$, then from (iv), $3r=5q=15k$, say, and we have the first form of automorph. If $p=-s$, then from (iv) $3r=-5q=-15k$, say, and we have the second form of automorph.

Historical note

The irrationality of square roots was known to the Pythagoreans (c. 550 B.C.) and the notion of a finite continued fraction is inherent in Euclid's division algorithm (c. 300 B.C.). The Indian mathematician Aryabhata used continued fractions for the solution of $ax+by=y$ about A.D. 500. Some periodic continued fractions for quadratic irrationals were known in the sixteenth century. John Wallis coined the phrase 'continued fraction' in a book published in 1655. In 1685 Wallis showed how continued fractions could be used to obtain good approximations to some irrationals. Most of the theory of continued fractions as we have used it appears in an edition of L. Euler's *Algebra* published in 1795 with additions by Lagrange.

In 1657, Fermat claimed that the equation $x^2-dy^2=1$ has an infinity of solutions when d is a non-square. L. Euler wrongly believed that a method given in Wallis' book was due to Fermat's contemporary John Pell, and the conventional name for the equation derives from this mistake. 'Fermat's equation' would be a more appropriate name. In 1765, Euler used the continued fraction for $\sqrt{d}$ to find a solution to $x^2-dy^2=1$, but it was Lagrange (in 1769, 1770) who determined the totality of solutions. The automorphs of indefinite forms were determined by G. P. L. Dirichlet's analysis of quadratic forms over the Gaussian integers in 1842.

11

Approximation of irrationals by rationals

Naive approach

1 What is the integer nearest to $\sqrt{2}$? What is the integer nearest to $\sqrt{3}$?

2 If α is a real number, what can be said about the value of $\alpha - [\alpha]$? Is there necessarily an integer n such that $|\alpha - n| \leqslant \frac{1}{2}$? If α is irrational, how can this inequality be improved?

3 If the integer points, n, and the points mid-way between them, $n + \frac{1}{2}$, are marked on a number line, how far away from the nearest of such points can any point on the number line be?
Find an integer m such that $|\sqrt{2} - \frac{1}{2}m| < \frac{1}{4}$, and an integer k such that $|\sqrt{5} - \frac{1}{2}k| < \frac{1}{4}$.
For any real number α, must there be an integer m such that $|\alpha - \frac{1}{2}m| < \frac{1}{4}$?

4 If for all integers n, the points n, $n + \frac{1}{3}$ and $n + \frac{2}{3}$ are marked on a number line, how far away from the nearest of such points can any point on the number line be?
Find an integer m such that $|\sqrt{2} - \frac{1}{3}m| < \frac{1}{6}$, and an integer k such that $|\sqrt{3} - \frac{1}{3}k| < \frac{1}{6}$.
For any irrational number α, must there be an integer m such that $|\alpha - \frac{1}{3}m| < \frac{1}{6}$?

5 For any irrational number α must there be an integer m such that $|\alpha - \frac{1}{3}m| < \frac{1}{9}$? What length of the closed interval $[1, 2]$ is within $\frac{1}{9}$ of one of the points $1, \frac{4}{3}, \frac{5}{3}$, or 2?
Find a real number α such that $|\alpha - \frac{1}{3}m| > \frac{1}{9}$ for all integers m.

6 If α is an irrational number and q is a given positive integer, does there always exist an integer p such that $|\alpha - p/q| < 1/2q$ and an integer p such that $|\alpha - p/q| < 1/q^2$?

7 Use a pocket calculator to determine the eleven numbers $n\sqrt{2} - [n\sqrt{2}]$ to two places of decimals for $n = 1, 2, \ldots, 11$.
Give reasons why no two of these eleven numbers are equal, and why none of them have a value $\frac{1}{10}k$ for any integer k.
Deduce that each of the eleven numbers belongs to exactly one of the open intervals $(0, \frac{1}{10})$, $(\frac{1}{10}, \frac{2}{10}), \ldots, (\frac{9}{10}, 1)$.
Select two of these numbers which lie in the same open interval and determine integers p and q such that $|q\sqrt{2} - p| < \frac{1}{10}$.

8 Use the method of q 7 to find integers p and q such that $|q\sqrt{3} - p| < \frac{1}{10}$. Does the method ensure that $q \leqslant 10$?

9 Generalise the method of q 7 and q 8 to show that for any irrational α and any predetermined positive integer n, it is possible to find integers p and q such that $|q\alpha - p| < 1/n$, with $0 < q \leqslant n$.

10 Deduce from q 9 that for any irrational α and any predetermined positive integer n, it is possible to find integers p and q such that $|\alpha - p/q| < 1/q^2$, with $0 < q \leqslant n$.

Farey sequences

11 To see how close the rational numbers with denominators $\leqslant n$ are, we explore the Farey sequence, F_n, as in q 8.13.

$$F_5 \quad \tfrac{0}{1} \quad \tfrac{1}{5} \quad \tfrac{1}{4} \quad \tfrac{1}{3} \quad \tfrac{2}{5} \quad \tfrac{1}{2} \quad \tfrac{3}{5} \quad \tfrac{2}{3} \quad \tfrac{3}{4} \quad \tfrac{4}{5} \quad \tfrac{1}{1}$$

$$F_6 \quad \tfrac{0}{1} \quad \tfrac{1}{6} \quad \tfrac{1}{5} \quad \tfrac{1}{4} \quad \tfrac{1}{3} \quad \tfrac{2}{5} \quad \tfrac{1}{2} \quad \tfrac{3}{5} \quad \tfrac{2}{3} \quad \tfrac{3}{4} \quad \tfrac{4}{5} \quad \tfrac{5}{6} \quad \tfrac{1}{1}$$

$$F_7 \quad \tfrac{0}{1} \quad \tfrac{1}{7} \quad \tfrac{1}{6} \quad \tfrac{1}{5} \quad \tfrac{1}{4} \quad \tfrac{2}{7} \quad \tfrac{1}{3} \quad \tfrac{2}{5} \quad \tfrac{3}{7} \quad \tfrac{1}{2} \quad \tfrac{4}{7} \quad \tfrac{3}{5} \quad \tfrac{2}{3} \quad \tfrac{5}{7} \quad \tfrac{3}{4} \quad \tfrac{4}{5} \quad \tfrac{5}{6} \quad \tfrac{6}{7} \quad \tfrac{1}{1}$$

Compare the number of terms in F_5 with $1 + \phi(1) + \phi(2) + \phi(3) + \phi(4) + \phi(5)$.
Conjecture the number of terms in F_n.
Justify your conjecture.

12 If a/b and c/d are adjacent terms in a Farey sequence, what is the area of the triangle with vertices $(0, 0)$, (b, a), and (d, c)? See q 8.12.

13 If a/b, c/d, e/f are adjacent terms, in order, of a Farey sequence and we let $O = (0, 0)$, $A = (b, a)$, $C = (d, c)$ and $E = (f, e)$, determine the areas of the triangles OCA and OCE. Let H and K be the feet of the perpendiculars from A and E to OC, respectively. Why should $AH = EK$?
What are the coordinates of the images of O, A and E under the

half-turn about the mid-point of HK? Deduce that $c/d = (a+e)/(b+f)$. Compare this method with q 8.14.

14 If a/b and c/d are adjacent terms in a Farey sequence, F_n, prove that the area of the triangles $(0, 0)$, (b, a), $(b+d, a+c)$ and $(0, 0)$, (d, c), $(b+d, a+c)$ are each $\frac{1}{2}$, and determine the shortest Farey sequence in which a/b, $(a+c)/(b+d)$, c/d are three adjacent terms in order.
Why must $b+d>n$?

15 If a, b, c, d are positive real numbers and $a/b<c/d$, give geometric reasons why

$$\frac{a}{b}<\frac{a+c}{b+d}<\frac{c}{d}$$

and why for any positive numbers u and v,

$$\frac{a}{b}<\frac{au+c}{bu+d}<\frac{c}{d} \quad \text{and} \quad \frac{a}{b}<\frac{a+cv}{b+dv}<\frac{c}{d}.$$

Compare your proof with q 10.16.

16 Give geometric or algebraic reasons why for any positive real numbers a, b, c, d, u, v such that

$$\frac{a}{b}<\frac{c}{d}, \frac{a}{b}<\frac{au+cv}{bu+dv}<\frac{c}{d}.$$

If a, b, c and d are positive integers, does every rational number between a/b and c/d have the form $(au+cv)/(bu+dv)$ for some suitable choice of integers u and v?

17 If a/b and c/d are adjacent terms in the Farey sequence F_n, $(a+c)/(b+d)$ is called the *mediant* of a/b and c/d. Prove that if the Farey sequences $F_n, F_{n+1}, F_{n+2}, \ldots$ are examined, the mediant is the first term to appear between a/b and c/d. Prove also that

$$\left|\frac{a+c}{b+d}-\frac{a}{b}\right| \leqslant \frac{1}{b(n+1)} \quad \text{and} \quad \left|\frac{c}{d}-\frac{a+c}{b+d}\right| \leqslant \frac{1}{d(n+1)}.$$

18 Which of the terms of F_2, F_3, F_4, F_5, F_6 and F_7 does $1/\sqrt{2}$ lie between?
Determine whether $1/\sqrt{2}$ is greater or less than the mediant in each case.
For $n=2, 3, 4, 5, 6, 7$, find a rational number a/b such that $|1/\sqrt{2}-a/b| \leqslant 1/b(n+1)$, where $b \leqslant n$.

19 If α is any real number between 0 and 1, must there exist integers a and b such that $|\alpha - a/b| \leqslant 1/8b$, where $b \leqslant 7$?

⟦239⟧

20 For a given positive integer n, and $0 \leqslant \alpha \leqslant 1$, explain why there exist integers a and b such that $|\alpha - a/b| \leqslant 1/b(n+1)$, where $b \leqslant n$.

21 Explain why the restriction $0 \leqslant \alpha \leqslant 1$ can be lifted without altering the validity of q 20.

22 Use q 20 to give an alternative proof of q 10.

Hurwitz' theorem

So far, we have established in two different ways that for any irrational α there exist integers p and q such that $|\alpha - p/q| < 1/q^2$. We prove that there are an infinity of rational numbers p/q satisfying this condition, and then see how the convergents to the continued fraction for α enables us to find them and to improve on the predicted closeness of the approximation.

23 For which of the positive integers $b = 1, 2, 3, 4$ and 5 does there exist an integer a such that $|\sqrt{2} - a/b| < 1/b^2$?

24 If α is the rational number p/q where $\gcd(p, q) = 1$, show that $|b\alpha - a| \geqslant 1/q$ for all non-zero integers a and b unless $a/b = p/q = \alpha$.

25 If $|\alpha - a/b| < 1/b^2$, and α is not rational, show that it is possible to find integers p and q such that $|\alpha - p/q| < 1/q^2$, with $|\alpha - p/q| < |\alpha - a/b|$, by a suitable choice of n in q 20. Deduce that there are an infinity of integral pairs a, b such that $|\alpha - a/b| < 1/b^2$.

26 If p_{n-1}/q_{n-1} and p_n/q_n are successive convergents to an irrational number α, use the fact that α lies between these convergents to prove that

$$\left|\alpha - \frac{p_{n-1}}{q_{n-1}}\right| < \frac{1}{q_{n-1}q_n} < \frac{1}{q_{n-1}^2}.$$

27 Use a calculator to find the values of $(n\sqrt{2} - [n\sqrt{2}])n$ to three decimal places for $n = 1, 2, \ldots, 30$.
Compare these values with the five values of $|\sqrt{2} - p/q|q^2$ as p/q takes on the values of the first five convergents to $\sqrt{2}$.
Conjecture a method of finding values of the integers p and q such that $|\sqrt{2} - p/q|q^2 < \frac{1}{2}$.

⟦240⟧

28 If p_n/q_n and p_{n+1}/q_{n+1} are consecutive convergents to $\sqrt{2}$ such that

$$\left|\sqrt{2}-\frac{p_n}{q_n}\right| q_n^2 \geq \tfrac{1}{2} \quad \text{and} \quad \left|\sqrt{2}-\frac{p_{n+1}}{q_{n+1}}\right| q_{n+1}^2 \geq \tfrac{1}{2}.$$

Use q 10.24 to show that

$$\frac{1}{q_n q_{n+1}} \geq \frac{1}{2}\left(\frac{1}{q_n^2}+\frac{1}{q_{n+1}^2}\right).$$

Deduce that $0 \geq (q_n - q_{n+1})^2$, which is absurd.
Establish that there are an infinity of rational numbers p/q, where p and q are coprime integers such that $|\sqrt{2}-p/q| < 1/2q^2$.

29 For any irrational α, show that there exist an infinity of coprime integers p and q such that $|\alpha - p/q| < 1/2q^2$, by showing that at least one of every consecutive pair of convergents to α satisfies the condition.

30 (i) If p_{n-1}/q_{n-1} and p_n/q_n are consecutive convergents to $\sqrt{2}$ such that

$$\left|\sqrt{2}-\frac{p_{n-1}}{q_{n-1}}\right| q_{n-1}^2 \geq \frac{1}{\sqrt{5}} \quad \text{and} \quad \left|\sqrt{2}-\frac{p_n}{q_n}\right| q_n^2 \geq \frac{1}{\sqrt{5}},$$

use q 10.24 to show that

$$\frac{1}{q_{n-1}q_n} \geq \frac{1}{\sqrt{5}}\left(\frac{1}{q_{n-1}^2}+\frac{1}{q_n^2}\right),$$

and deduce that

$$\sqrt{5} \geq \frac{q_n}{q_{n-1}}+\frac{q_{n-1}}{q_n}.$$

Why is equality impossible here?

(ii) Sketch the graph of $y = x + 1/x$ for positive x. What values can x have if $y < \sqrt{5}$? If $x > 1$ and $y < \sqrt{5}$, show that $x < \frac{1}{2}(\sqrt{5}+1)$ and $1/x > \frac{1}{2}(\sqrt{5}-1)$.

(iii) If p_{n-1}/q_{n-1}, p_n/q_n and p_{n+1}/q_{n+1} are consecutive convergents to $\sqrt{2}$ such that $|\sqrt{2}-p_i/q_i|q_i^2 \geq 1/\sqrt{5}$ for $i = n-1, n, n+1$, use (i) and (ii) to show that

$$1+\frac{q_{n-1}}{q_n} > \tfrac{1}{2}(\sqrt{5}+1) > \frac{q_{n+1}}{q_n}.$$

Use the relationship between q_{n-1}, q_n and q_{n+1} of q 10.13 to obtain a contradiction.

⟦240⟧

31 For any irrational number α, show that there are an infinity of pairs of coprime integers p and q such that $|\alpha - p/q| < 1/\sqrt{5}q^2$ by showing that at least one of every three consecutive convergents to α satisfies this condition. (*Hurwitz' theorem*)

In the next two questions we establish that the $\sqrt{5}$ which appears in the statement of Hurwitz' theorem is the greatest possible number which can appear consistent with the validity of such a theorem.

32 Determine the continued fraction for $\frac{1}{2}(\sqrt{5}+1)$.

33 (i) If $|\frac{1}{2}(\sqrt{5}+1) - p/q| = 1/cq^2$, prove that $p^2 - pq - q^2 = 1/c^2q^2 \pm \sqrt{5}/c$.
(ii) For non zero integers p and q prove that $p^2 - pq - q^2 \neq 0$.
(iii) If $c > \sqrt{5}$, prove that for sufficiently large q, $-1 < 1/c^2q^2 \pm \sqrt{5}/c < 1$.
(iv) Deduce from (i), (ii) and (iii), that when $c > \sqrt{5}$, there are at most a finite number of rationals p/q such that $|\frac{1}{2}(\sqrt{5}+1) - p/q| < 1/cq^2$.

Liouville's theorem

Up to this point, we have been trying to find good rational approximations to irrational numbers and to claim the degree of closeness which may be obtained when the convergents to the relevant continued fraction are used. In the rest of the chapter, we shall find that for a particular class of irrational numbers (the algebraic numbers) even the best rational approximations are not particularly close.

34 If p and q are positive integers, explain why $|p^2/q^2 - 2| \geqslant 1/q^2$. Sketch the graph of $y = x^2 - 2$ and explain why

$$\frac{|p^2/q^2 - 2|}{|p/q - \sqrt{2}|} \leqslant \frac{2}{2-\sqrt{2}}, \quad \text{when } 1 \leqslant p/q \leqslant 2.$$

Deduce that $|p/q - \sqrt{2}| \geqslant (2-\sqrt{2})/2q^2$.
Explain why this inequality holds even when p/q does not lie between 1 and 2. Prove that for all non-zero integers p and q, $|p/q - \sqrt{2}| > 1/4q^2$.

35 For what positive integers p and q is $|p/q - \sqrt{2}| < 1/q^3$?
For any positive real number c, explain why there are at most a finite number of coprime pairs p and q such that $|p/q - \sqrt{2}| < c/q^3$.

⟦242⟧

36 Sketch the graph of $y=x^2-3$, and using the method of q 34, prove that for positive integers p and q $|p/q-\sqrt{3}|\geqslant(2-\sqrt{3})/q^2>1/5q^2$. Explain why for any positive real number c, there are at most a finite number of coprime pairs p, q such that $|p/q-\sqrt{3}|<c/q^3$.

37 Sketch the graph of $y=x^2-x-1$, and prove that for positive integers $p,\ q, |p^2/q^2-p/q-1|\geqslant 1/q^2$.
Show that

$$\frac{|p^2/q^2-p/q-1|}{|p/q-\frac{1}{2}(\sqrt{5}+1)|}\leqslant\frac{2}{3-\sqrt{5}}.$$

Deduce that $|p/q-\frac{1}{2}(\sqrt{5}+1)|\geqslant(3-\sqrt{5})/2q^2>1/3q^2$.

38 Prove that $\sqrt[3]{2}$ is not rational. Deduce that for any positive integers p and q, $|p^3/q^3-2|\geqslant 1/q^3$.
Sketch the graph of $y=x^3-2$ and show that

$$\frac{|p^3/q^3-2|}{|p/q-\sqrt[3]{2}|}\leqslant\frac{6}{2-\sqrt[3]{2}},\quad \text{when } 1\leqslant p/q\leqslant 2.$$

Deduce that $|p/q-\sqrt[3]{2}|\geqslant(2-\sqrt[3]{2})/6q^3>1/12q^3$.
Explain why $|p/q-\sqrt[3]{2}|\geqslant 1/12q^3$ for all positive integers p and q.
Prove that there are at most a finite number of coprime pairs p, q such that $|p/q-\sqrt[3]{2}|<1/q^4$ and that there are at most a finite number of coprime pairs p, q such that $|p/q-\sqrt[3]{2}|<c/q^4$ for any given positive real number c.

39 Prove that the number $\frac{1}{3}(2-\sqrt[3]{2})$ is not rational. Prove further that it is a root of the equation $9x^3-18x^2+12x-2=0$ and in fact its only real root.
From the fact that this equation has no rational roots deduce that for any positive integers p, q

$$\left|9\frac{p^3}{q^3}-18\frac{p^2}{q^2}+12\frac{p}{q}-2\right|\geqslant\frac{1}{q^3}.$$

Draw the graph of $y=9x^3-18x^2+12x-2$ for $0<x<1$ as carefully as you can.
Deduce that

$$\frac{|9p^3/q^3-18p^2/q^2+12p/q-2|}{|p/q-\frac{1}{3}(2-\sqrt[3]{2})|}\leqslant\frac{6}{2-\sqrt[3]{2}}$$

when $0<p/q<1$ and that $|p/q-\frac{1}{3}(2-\sqrt[3]{2})|\geqslant(2-\sqrt[3]{2})/6q^3$. Explain why $|p/q-\frac{1}{3}(2-\sqrt[3]{2})|>1/12q^3$ for all positive integers p, q and deduce that there are at most a finite number of coprime pairs p, q such that $|p/q-\frac{1}{3}(2-\sqrt[3]{2})|<c/q^4$ for a given positive number c.

⟦242⟧

40 If α is an irrational number which satisfies an equation of the form

$$a_n x^n + a_{n-1} x^{n-1} + \ldots + a_1 x + a_0 = 0,$$

where the coefficients are integers and $a_n \neq 0$, and α satisfies no such equation of lower degree, prove that for any positive integer q and any integer p

$$\left| a_n \frac{p^n}{q^n} + a_{n-1} \frac{p^{n-1}}{q^{n-1}} + \ldots + a_1 \frac{p}{q} + a_0 \right| \geq \frac{1}{q^n}.$$

Such an α is said to be an *algebraic number of degree n.*

41 If f is a real differentiable function with a root α, the mean value theorem asserts that for any real number $b \neq \alpha$, $f(b)/(b-\alpha)$ is equal to $f'(x)$ for some x, $\alpha \leq x \leq b$ or $b \leq x \leq \alpha$. If the maximum value of $|f'(x)|$ in the closed interval $[\alpha - \frac{1}{2}, \alpha + \frac{1}{2}]$ is A, deduce that $|b-\alpha| \geq |f(b)|/A$ when $\alpha - \frac{1}{2} \leq b \leq \alpha + \frac{1}{2}$.

42 If α is an algebraic number of degree $n \geq 2$, prove that there exists a positive real number A such that $|p/q - \alpha| > 1/Aq^n$ for all integers p and all positive integers q. (*Liouville's theorem*)

43 If $\alpha = \dfrac{1}{10^{1!}} + \dfrac{1}{10^{2!}} + \dfrac{1}{10^{3!}} + \dfrac{1}{10^{4!}} + \ldots$,

that is

$$\alpha = \frac{1}{10} + \frac{1}{10^2} + \frac{1}{10^6} + \frac{1}{10^{24}} + \ldots$$

and

$$\frac{p_1}{q_1} = \frac{1}{10}, \frac{p_2}{q_2} = \frac{1}{10} + \frac{1}{10^2}, \frac{p_3}{q_3} = \frac{1}{10} + \frac{1}{10^2} + \frac{1}{10^6},$$

and in general

$$\frac{p_n}{q_n} = \frac{1}{10^{1!}} + \frac{1}{10^{2!}} + \frac{1}{10^{3!}} + \ldots + \frac{1}{10^{n!}},$$

where p_n and q_n are coprime, find $p_1, q_1, p_2, q_2, p_3, q_3, q_n$. Prove that

$$0 < \alpha - \frac{p_1}{q_1} < \frac{1}{10^2} + \frac{1}{10^3} + \frac{1}{10^4} + \ldots + \frac{1}{10^i} + \ldots = \frac{1}{10^2}\left(\frac{10}{9}\right) = \frac{1}{9}\frac{1}{10} < \frac{1}{q_1},$$

$$0 < \alpha - \frac{p_2}{q_2} < \frac{1}{10^6} + \frac{1}{10^7} + \frac{1}{10^8} + \ldots = \frac{1}{10^6}\left(\frac{10}{9}\right) = \frac{1}{9}\frac{1}{10^5} < \left(\frac{1}{q_2}\right)^2,$$

$$0 < \alpha - \frac{p_3}{q_3} < \frac{1}{10^{24}} + \frac{1}{10^{25}} + \frac{1}{10^{26}} + \ldots = \frac{1}{10^{24}}\left(\frac{10}{9}\right) = \frac{1}{9}\frac{1}{10^{23}} < \left(\frac{1}{q_3}\right)^3,$$

and finally

$$0 < \alpha - \frac{p_n}{q_n} < \frac{1}{10^{(n+1)!}} + \frac{1}{10^{(n+1)!+1}} + \ldots = \frac{1}{10^{(n+1)!}}\left(\frac{10}{9}\right) < \left(\frac{1}{q_n}\right)^n.$$

Deduce that for no value of n or A can α satisfy the condition of q 42, so α is not an algebraic number.

A real number which is not algebraic is said to be *transcendental.*

Notes and answers

For concurrent reading, see Niven (1961).

1 $\sqrt{2}-1<\frac{1}{2}$, $2-\sqrt{3}<\frac{1}{2}$.

2 $0\leqslant\alpha-[\alpha]<1$. If $\frac{1}{2}<\alpha-[\alpha]$ then $[\alpha]+1-\alpha<\frac{1}{2}$. If α is irrational $\alpha\neq\frac{1}{2}n$ for any integer and there is an integer n such that $|\alpha-n|<\frac{1}{2}$.

3 $\frac{1}{4}$, $|\sqrt{2}-\frac{3}{2}|<\frac{1}{4}$, $|\sqrt{5}-2|<\frac{1}{4}$. Unless $\alpha=n/4$ for some odd integer n.

4 $\frac{1}{6}$, $|\sqrt{2}-\frac{4}{3}|<\frac{1}{6}$, $|\sqrt{3}-\frac{5}{3}|<\frac{1}{6}$.
Yes, since $\alpha\neq n/6$ for any integer n.

5 Not if $\frac{1}{9}<\alpha<\frac{2}{9}$ for example. $\frac{2}{3}$. $\frac{3}{18}$, for example.

6 For $|\alpha-p/q|<1/2q$, p may always be found since α lies in some interval of the kind $(m/q,(m+1)/q)$ and is therefore $<1/2q$ from one or other of the ends. If $q>2$, then

$$\left|\frac{2m+1}{2q}-\frac{m}{q}\right|=\frac{1}{2q}>\frac{1}{q^2}.$$

7 If $m\neq n$, but $n\sqrt{2}-[n\sqrt{2}]=m\sqrt{2}-[m\sqrt{2}]$, then $\sqrt{2}=([n\sqrt{2}]-[m\sqrt{2}])/(n-m)$ and $\sqrt{2}$ would be rational. Likewise if $n\sqrt{2}-[n\sqrt{2}]=k/10$.
Both $\sqrt{2}-[\sqrt{2}]$ and $6\sqrt{2}-[6\sqrt{2}]$ lie between 0.4 and 0.5, so $|(6\sqrt{2}-8)-(\sqrt{2}-1)|<\frac{1}{10}$, and $|5\sqrt{2}-7|<\frac{1}{10}$.
Both $2\sqrt{2}-[2\sqrt{2}]$ and $7\sqrt{2}-[7\sqrt{2}]$ lie between 0.8 and 0.9, so $|(7\sqrt{2}-9)-(2\sqrt{2}-2)|<\frac{1}{10}$, and $|5\sqrt{2}-7|<\frac{1}{10}$.

8 $|4\sqrt{3}-7|<\frac{1}{10}$. q is the difference of the two numbers $\leqslant 11$.

9 The $n+1$ numbers $i\alpha-[i\alpha]$ for $i=1,\ldots,n+1$ lie within the n open intervals $((i-1)/n, i/n)$ for $i=1,\ldots,n$. So two lie in the same interval, $i=j,k$ (say). Then $|j\alpha-[j\alpha]-k\alpha+[k\alpha]|<1/n$, and we have our result.

10 In q 9, $0<j,k\leqslant n+1$, so $0<|j-k|\leqslant n$, and so there exist integers p,q, such that $|q\alpha-p|<1/n$ for $q\leqslant n$. Now $|\alpha-p/q|<1/nq<1/q^2$.

11 The terms in F_{n+1} which were not in F_n are precisely those $k/(n+1)$, where $\gcd(k,n+1)=1$. There are just $\phi(n+1)$ of these, so F_{n+1} has $\phi(n+1)$ more terms than F_n and this gives the number of terms in F_n by induction.

12 $\frac{1}{2}$.

13 Area of OCA = area of $OCE=\frac{1}{2}$.
If we consider these two triangles as on the base OC, the altitudes from A and E to this base are equal. If AE meets OC at M, the triangles AMH and EMK are congruent, so the mid-point of HK is the mid-point of AE. The half-turn about $M=(\frac{1}{2}(b+f),\frac{1}{2}(a+e))$ interchanges A

and E, maps O to $(b+f, a+e)$ and, since M lies on OC, $(a+e)/(b+f)=c/d$.

14 The area of the triangle $(0, 0)$, (b, a), (d, c) is $\frac{1}{2}$, so the area of the parallelogram $(0, 0)$, (b, a), $(b+d, a+c)$, (d, c) is 1 and the two triangles of the question each occupy one half of this parallelogram. The three terms are adjacent in the Farey sequence F_{b+d}, and since there is no lattice point on the line segment joining $(0, 0)$, to $(b+d, a+c)$ they are not all three in any earlier Farey sequence.
If $b+d \leqslant n$, a/b and c/d would not be adjacent terms in F_n.

15 The line joining $(0, 0)$ to $(b+d, a+c)$ is the diagonal of the parallelogram of which $(0, 0)(b, a)$ and $(0, 0)(d, c)$ are sides. The line joining $(0, 0)$ to $(bu+d, au+c)$ is the diagonal of the parallelogram of which $(0, 0)(bu, au)$ and $(0, 0)(d, c)$ are sides.

16 For any real numbers r and s, the number $kr+(1-k)s$ divides them in the ratio $(1-k)/k$ for $0<k<1$, and so every number between these two is of this form. See q 9.31.
Now

$$\frac{a}{b}\frac{bu}{bu+dv}+\frac{c}{d}\left(1-\frac{bu}{bu+dv}\right)=\frac{au+cv}{bu+dv} \quad \text{and} \quad 0<\frac{bu}{bu+dv}<1$$

since all the terms are positive.
If r and s are rational numbers, the rational numbers between them have the form $kr+(1-k)s$ for rational k such that $0<k<1$. If such a k is given and b and d are given integers, one need only choose u and v as integers such that $v/u=b/dk-b/d$ to show that every rational number between a/b and c/d has the given form.

17 If a/b and c/d are adjacent terms in a Farey sequence, the area of the triangle $(0, 0)$, (b, a), (d, c) is $\frac{1}{2}$ from 12. If p/q is adjacent to a/b in some other Farey sequence $(0, 0)$, (b, a), (q, p) has area $\frac{1}{2}$, so that if c/d and p/q are both greater than or both less than a/b, the line joining (d, c) to (q, p) is parallel to the line $(0, 0)(b, a)$. There is no lattice point between (d, c) and $(b+d, a+c)$ so $(a+c)/(b+d)$ is the first term to appear in a Farey sequence between a/b and c/d.

$$\left|\frac{a+c}{b+d}-\frac{a}{b}\right|=\left|\frac{bc-ad}{b(b+d)}\right|=\frac{1}{b(b+d)}\leqslant\frac{1}{b(n+1)}.$$

18 $\frac{1}{2}<1/\sqrt{2}<1$, $\frac{2}{3}<1/\sqrt{2}<1$, $\frac{2}{3}<1/\sqrt{2}<\frac{3}{4}$, $\frac{2}{3}<1/\sqrt{2}<\frac{5}{7}$.
F_2: $\frac{1}{2}<1/\sqrt{2}<1$, above mediant. $a/b=1$.
F_3: $\frac{2}{3}<1/\sqrt{2}<1$, below mediant. $a/b=\frac{2}{3}$.
F_4, F_5, F_6: $\frac{2}{3}<1/\sqrt{2}<\frac{3}{4}$, below mediant. $a/b=\frac{2}{3}$.
F_7: $\frac{2}{3}<1/\sqrt{2}<\frac{5}{7}$, above mediant. $a/b=\frac{5}{7}$.

19 Yes, by q 17, since every number a/b between 0 and 1 is within $1/8b$ of a term of the Farey sequence F_7.

20 Every number a/b between 0 and 1 is within $1/b(n+1)$ of a term of the Farey sequence F_n.

21 For any real number α, there exist integers a and b such that

$$\left|\alpha-[\alpha]-\frac{a}{b}\right|\leqslant\frac{1}{b(n+1)},\quad \text{with } b\leqslant n,$$

so

$$\left|\alpha-\frac{b[\alpha]+a}{b}\right|\leqslant\frac{1}{b(n+1)}.$$

23 1, 2, 3 and 5.

24 $\left|\frac{bp}{q}-a\right|=\left|\frac{bp-qa}{q}\right|\geqslant\frac{1}{q}$ unless $bp=qa$.

25 Since α is irrational, $|\alpha-a/b|>0$, so for some integer n, $|\alpha-a/b|>1/n$. By q 20, there exist integers p, q such that

$|\alpha-p/q|\leqslant 1/q(n+1)<1/n<|\alpha-a/b|$

with $q\leqslant n$. Now $|\alpha-p/q|<1/q^2$ and the process may be repeated to give an infinity of such pairs.

26 $\left|\frac{p_n}{q_n}-\frac{p_{n-1}}{q_{n-1}}\right|=\left|\frac{p_nq_{n-1}-p_{n-1}q_n}{q_nq_{n-1}}\right|=\frac{1}{q_nq_{n-1}},$

and

$$\left|\frac{p_n}{q_n}-\frac{p_{n-1}}{q_{n-1}}\right|>\left|\alpha-\frac{p_{n-1}}{q_{n-1}}\right|.$$

27

0.414	6.120	14.668
1.657	11.647	2.479
0.728	5.002	12.119
2.627	11.186	22.587
0.355	3.198	8.883
2.912	10.039	20.008
6.296	0.708	4.962
2.510	8.205	16.743
6.551	16.531	0.354
1.421	5.685	12.792

$\sqrt{2}=[1,\dot{2}]$. First five convergents are $\frac{1}{1}, \frac{3}{2}, \frac{7}{5}, \frac{17}{12}, \frac{41}{29}$.
$|\sqrt{2}-p/q|q^2=0.414, 0.343, 0.355, 0.353, 0.354$.

28 Because $\sqrt{2}$ lies between its consecutive convergents

$$\left|\sqrt{2}-\frac{p_n}{q_n}\right|+\left|\sqrt{2}-\frac{p_{n+1}}{q_{n+1}}\right|=\left|\frac{p_n}{q_n}-\frac{p_{n+1}}{q_{n+1}}\right|.$$

The argument shows that at least one of every two consecutive convergents satisfies the condition $|\sqrt{2}-p/q|<1/2q^2$.

29 The argument of q 28 holds if any irrational α replaces $\sqrt{2}$.

30 (i) The result stems from the fact that $\sqrt{2}$ lies between its consecutive convergents as in q 28. Since q_{n-1} and q_n are integers, equality is impossible.

(ii) $y = x + 1/x$ is a hyperbola with $x = 0$ and $y = x$ as asymptotes. If $x + 1/x = \sqrt{5}$, we have $x = \frac{1}{2}(\sqrt{5} \pm 1)$ and $x + 1/x < \sqrt{5}$ between these values. Now $\frac{1}{2}(\sqrt{5} - 1) < 1 < \frac{1}{2}(\sqrt{5} + 1)$, so if $x > 1$ and $y < \sqrt{5}$, $x < \frac{1}{2}(\sqrt{5} + 1)$. This is equivalent to $1/x > \frac{1}{2}(\sqrt{5} - 1)$.

(iii) From (i) we have

$$\sqrt{5} > \frac{q_n}{q_{n-1}} + \frac{q_{n-1}}{q_n} \quad \text{and} \quad \sqrt{5} > \frac{q_{n+1}}{q_n} + \frac{q_n}{q_{n+1}}.$$

Now $q_{n+1} > q_n > q_{n-1} \geqslant 1$, so from (ii)

$$\frac{q_{n-1}}{q_n} > \tfrac{1}{2}(\sqrt{5} - 1) \quad \text{and} \quad \frac{q_{n+1}}{q_n} < \tfrac{1}{2}(\sqrt{5} + 1).$$

Thus $1 + q_{n-1}/q_n > q_{n+1}/q_n$ and $q_n + q_{n-1} > q_{n+1}$. But $q_{n+1} = a_n q_n + q_{n-1}$ where $a_n \geqslant 1$, so this is absurd.

31 The argument of q 30 holds with α for $\sqrt{2}$.

32 $[1, 1, 1, \dot{1}]$.

33 (i) $|\frac{1}{2}(\sqrt{5} + 1) - p/q| = 1/cq^2$

$$\Rightarrow \tfrac{1}{2}\sqrt{5}q \pm 1/cq = p - \tfrac{1}{2}q$$
$$\Rightarrow 5q^2/4 \pm \sqrt{5}/c + 1/c^2q^2 = p^2 - pq + q^2/4$$
$$\Rightarrow 1/c^2q^2 \pm \sqrt{5}/c = p^2 - pq - q^2.$$

(ii) If $p^2 - pq - q^2 = 0$, then $(p/q)^2 - (p/q) - 1 = 0$, so $p/q = \frac{1}{2}(1 \pm \sqrt{5})$, which contradicts the irrationality of $\sqrt{5}$.

(iii) If $c > \sqrt{5}$, by choosing $q^2 > \frac{1}{5}(1 - \sqrt{5}/c)$,

$$-1 < 1/c^2q^2 \pm \sqrt{5}/c < 1.$$

(iv) If $c > \sqrt{5}$, then from (iii) and (i), for all but finitely many q, $-1 < p^2 - pq - q^2 < 1$. But p and q are integers so $p^2 - pq - q^2$ is an integer, so from (ii), this condition is never satisfied.

It is instructive to find the c of part (i) when p/q is one of the first ten convergents to $\frac{1}{2}(\sqrt{5} + 1)$.

This argument may be carried through for approximations to any real number for which the continued fraction expansion is ultimately a sequence of 1s.

34 Since $\sqrt{2}$ is irrational, there are no integers p, q such that $p^2 - 2q^2 = 0$. Thus

$$|p^2/q^2 - 2| = |(p^2 - 2q^2)/q^2| \geqslant 1/q^2.$$

The slope of the chord joining (x, x^2-2) to $(\sqrt{2}, 0)$ is greatest for $1 \leqslant x \leqslant 2$ when $x=2$.
For any rational number $s/q \neq 0$, there is a p/q between 1 and 2 such that

$|s/q-\sqrt{2}| \geqslant |p/q-\sqrt{2}| \geqslant (2-\sqrt{2})/2q^2$.

$9>8 \Rightarrow 3>2\sqrt{2} \Rightarrow 4-2\sqrt{2}>1 \Rightarrow 2-\sqrt{2}>\frac{1}{2}$.

We can conclude that if every rational point p/q is removed from the real line together with a neighbourhood of radius $1/4q^2$ about that point, then the number $\sqrt{2}$ will not be removed.

35 From q 34, $q<4$. If $q=1$, $p=1$ or 2. If $q=2$, $p=3$. $|\frac{4}{3}-\sqrt{2}|$ and $|\frac{5}{3}-\sqrt{2}|>\frac{1}{27}$.
From q 34, $q/c<4$, so at most a finite number of q are admissible. If $n=[\sqrt{2}+c/q^3]+1$, only ps between $\pm nq$ need be considered.

36 The slope of the chord joining (x, x^2-3) to $(\sqrt{3}, 0)$ is greatest for $1 \leqslant x \leqslant 2$ when $x=2$. Together with $|p^2/q^2-3| \geqslant 1/q^2$, this gives $|p/q-\sqrt{3}| \geqslant (2-\sqrt{3})/q^2$.
Now $81>75 \Rightarrow 9>5\sqrt{3} \Rightarrow 10-5\sqrt{3}>1 \Rightarrow 2-\sqrt{3}>\frac{1}{5}$.
From this argument $c/q^3>|p/q-\sqrt{3}|>1/5q^2 \Rightarrow 5c>q$, so only a finite number of q need to be considered.

37 The slope of the chord joining (x, x^2-x-1) to $(\frac{1}{2}(\sqrt{5}+1), 0)$ is greatest for $1 \leqslant x \leqslant 2$ when $x=2$ and thus less than $1/[2-\frac{1}{2}(\sqrt{5}+1)]$.
$49>45 \Rightarrow 7>3\sqrt{5} \Rightarrow 9-3\sqrt{5}>2 \Rightarrow \frac{1}{2}(3-\sqrt{5})>\frac{1}{3}$.
Compare this result with q 33.

38 If $\sqrt[3]{2}$ were a rational number $=p_1^{a_1}p_2^{a_2}\ldots p_n^{a_n}$ in its prime factorisation (q 10.5) then

$2=p_1^{3a_1}p_2^{3a_2}\ldots p_n^{3a_n}$,

which is impossible. Thus for non-zero integers p, q, $p^3-2q^3 \neq 0$ and $|p^3/q^3-2| \geqslant 1/q^3$.
The graph of $y=x^3-2$ is concave upwards for $1<x<2$ and so the slope of the line joining $(\sqrt[3]{2}, 0)$ to (x, x^3-2) in this interval is greatest when $x=2$, and thus less than or equal to $6/(2-\sqrt[3]{2})$.

$27>16 \Rightarrow 3>2\sqrt[3]{2} \Rightarrow 4-2\sqrt[3]{2}>1 \Rightarrow 24-12\sqrt[3]{2}>6 \Rightarrow \frac{1}{6}(2-\sqrt[3]{2})>\frac{1}{12}$.

The argument continues as in q 35.

39 If $\frac{1}{3}(2-\sqrt[3]{2})=r$ were rational, then $\sqrt[3]{2}=2-3r$ would be rational, and this contradicts q 38.
$(2-3r)^3=2$ so $-27r^3+54r^2-36r+8=2$, and $9r^3-18r^2+12r-2=0$. If we consider the graph of $y=9x^3-18x^2+12x-2$,
$dy/dx=27x^2-36x+12=3(3x-2)^2$.

Since the gradient is always positive or zero, the curve cuts the x-axis exactly once, so no rational number is a root and $9p^3/q^3 - 18p^2/q^2 + 12p/q - 2$ is not zero.

The graph is concave downwards for $0 < x < \frac{2}{3}$, and concave upwards for $\frac{2}{3} < x < 1$. The root is near 0.25. Thus for $0 < x < 1$, the chord with one end at $(\frac{1}{3}(2 - \sqrt[3]{2}), 0)$ and greatest slope has $(0, -2)$ at the other extremity. This has the same slope as the chord in q 38 so the arithmetic is similar.

40 The number concerned is clearly an integer divided by q^n, so we only need to establish that the number is not zero. If it was zero, the equation would have a rational root p/q and the polynomial would have a factor $(qx - p)$ so that α would satisfy an equation of degree $n - 1$.

41 $A \geqslant |f'(x)| \geqslant |f(b)/(b - \alpha)|$.

42 For any positive integers p and q there is an integer p' such that $\alpha - \frac{1}{2} \leqslant p'/q \leqslant \alpha + \frac{1}{2}$, and if p/q is not in this interval,

$|p/q - \alpha| > |p'/q - \alpha|$,

so it is sufficient to consider p/q within the interval $[\alpha - \frac{1}{2}, \alpha + \frac{1}{2}]$. Take A as the greatest value of $|f'(x)|$ in this interval, where $f(x)$ is a polynomial with integer coefficients such that $f(\alpha) = 0$. Apply q 41 and then q 40.

Historical note

The argument of q 9 uses P. G. L. Dirichlet's 'pigeon hole' principle that if $n + 1$ elements are distributed among n sets, at least one set contains two or more of these elements. When J. Hurwitz proved his theorem about rational approximations to irrational numbers in 1891, he used Farey sequences, not continued fractions, to do it. In 1903, E. Borel showed that at least one out of every three consecutive convergents of the continued fraction for an irrational number satisfies the condition of Hurwitz theorem. The proof of this that we have developed is due to O. Perron (1910). It was in 1851 that J. Liouville constructed the first transcendental numbers which could be proved to be such. In 1873, C. Hermite proved the transcendence of e, and in 1882, F. Lindemann proved the transcendence of π.

BIBLIOGRAPHY

Andrews, G. E., 1971, *Number theory*, Saunders.
Chapters 12 and 13, on partitions, may be read concurrently with our chapter 7.

Bell, E. T., 1962, *The last problem*, Gollancz.
Contains an interesting selection from Fermat's letters.

Bolker, E. D., 1970, *Elementary number theory*, Benjamin.
Uses groups, rings and fields; may be read concurrently with our chapters 4, 5 and 6.

Butts, T., 1973, *Problem solving in mathematics*, Scott-Foresman.
An initiation, suitable for preliminary work, or for working concurrently with our chapter 1.

Chrystal, G., 1964, *Algebra* vol. 2, Chelsea.
Contains sections which may be read concurrently with our chapters 7 and 10.

Davenport, H., 1968, *The higher arithmetic*, Hutchinson.
Strongly recommended for concurrent reading throughout.

Dickson, L. E., 1950, *History of the theory of numbers*, 3 vols., Chelsea.
Extensive and detailed references.

Edwards, H. M., 1977, *Fermat's last theorem, a genetic introduction to algebraic number theory*, Springer.
Discusses substantial mathematics in a readable historical sequence; will be appreciated best after completing the *Pathway*, but may be started concurrently with our chapters 5 and 6.

Hardy, G. H. & Wright, E. M., 1980, *An introduction to the theory of numbers*, Oxford, 5th edn.
The classical reference book on undergraduate number theory; difficult for beginners to learn from, except for chapter 3 which may be read concurrently with our chapter 9.

Hubbard, R. L., 1975, *The factor book*, Hilton Management Services.
Gives the prime factors of numbers up to 100 000.

LeVeque, W. J., 1956, *Topics in number theory*, Addison-Wesley.
Vol. 1 suitable for further reading on arithmetical functions and the distribution of primes; vol. 2 suitable for further reading on quadratic forms.

Mathews, G. B. (no date), *Number theory*, Chelsea.
Contains several proofs of the law of quadratic reciprocity and further reading on quadratic forms.

Nagell, T., 1964, *Introduction to number theory*, Chelsea.
Gives three different forms of Legendre's theorem.

Niven, I., 1961, *Numbers rational and irrational*, Mathematical Association of America.
An introduction to our chapter 11 for a beginner, finishing with the same construction of a transcendental number as we do.

Niven, I., 1967, *Irrational numbers*, Carus series, Mathematical Association of America.
Further reading on our chapter 11.

Niven I. & Zuckerman, H. S., 1972, *An introduction to the theory of numbers*, Wiley, 3rd edn.
A good conventional textbook, with exercises, covering most of the *Pathway*.

Olds, C. D., 1963, *Continued fractions*, Mathematical Association of America.
Strongly recommended for concurrent reading with our chapter 10.

Ore, O., 1948, *Number theory and its history*, McGraw-Hill.
May be read concurrently with our chapters 2, 3 and 6.

Ore, O., 1967, *Invitation to number theory*, Mathematical Association of America.
May be read concurrently with our chapter 1.

Pollard, H. & Diamond, H. G., 1975, *The theory of algebraic numbers*, Carus series, Mathematical Association of America, 2nd edn.
Further reading on our chapter 5.

Pólya, G. & Szegö, G., 1976, *Problems and theorems in analysis*, vol. 2, Springer.
Part 8 pursues more advanced number theory in a style similar to that of the *Pathway*.

Reid, C., 1956, *From zero to infinity*, Routledge and Kegan Paul.
Gives an authentic taste of number theory to beginners about to start on the *Pathway*.

Roberts, J., 1977, *Elementary number theory, a problem oriented approach*, M.I.T.
Goes further and faster than the *Pathway* using a sequence of questions.

Shanks, D., 1978, *Solved and unsolved problems in number theory*, Chelsea, 2nd edn.
May be read concurrently with our chapters 3, 4, 5, 6 and 10.

Sierpinski, W., 1962, *Pythagorean triangles*, Graduate School of Science, Yeshiva University, New York.
May be read concurrently with the first part of our chapter 5.

Stark, H. M., 1978, *An introduction to number theory*, M.I.T.
Chapter 7 uses a geometric approach to continued fractions originally suggested by F. Klein.

Weil, A., 1979, *Number theory for beginners*, Springer.
A concise treatment, using the terms of modern algebra, which may be read concurrently with our chapter 2.

DEFINITIONS AND THEOREMS

The large bold figures represent the chapter numbers

1

Division algorithm (1.24)
If a and b are non-zero integers, then there exist unique integers q, r such that $a = bq + r$, where $0 \leqslant r < |b|$.

Definition: greatest common divisor (1.34)
If a and b are non-zero integers, and d is the greatest positive integer such that $d|a$ and $d|b$, then $d = \gcd(a, b)$.

Euclidean algorithm (1.34, 1.39)
If $d = \gcd(a, b)$, then there exist integers x and y such that $d = ax + by$.

Definition: prime number (1.43)
If p is a given positive integer different from 1, and $a|p$ implies $a = \pm 1, \pm p$, then p is said to be prime.

Fundamental theorem of arithmetic (1.58)
Each positive integer greater than 1 equals $p_1^{\alpha_1} p_2^{\alpha_2} \ldots p_k^{\alpha_k}$, where the p_i form a unique set of distinct prime numbers, the α_i are unique positive integers and $k \geqslant 1$.

Theorem (1.64)
There exists an infinity of prime numbers.

2

Definition: residue class (2.3)
The set of integers $\{x : n|x - a\}$ is said to form a residue class modulo n. The set of n residue classes modulo n is denoted by $\mathbf{Z}_n$. We write $x \equiv y \pmod{n}$ when x and y lie in the same residue class modulo n.

Chinese remainder theorem (2.18)
If $m_1, m_2, \ldots, m_n$ are coprime in pairs, then there is a unique solution modulo $m_1m_2\ldots m_n$ to the n congruences

$x \equiv a_i \pmod{m_i}$, $i = 1, 2, \ldots, n$.

Definition: modular addition (2.26)
If the integer a belongs to the residue class $[a]$ modulo n, and the integer b belongs to the residue class $[b]$ modulo n, then addition modulo n is defined on $\mathbf{Z}_n$ so that $[a]+[b]=[a+b]$.

Definition: Euler's ϕ function (2.39)
The number of generators of the group $(\mathbf{Z}_n, +)$ is denoted by $\phi(n)$. Equivalently, $\phi(n)$ is the number of numbers between 1 and n inclusive which are coprime to n.

Theorem: ϕ is multiplicative (2.50)
If m and n are coprime, $\phi(mn)=\phi(m)\phi(n)$.

Theorem: the value of $\phi(n)$. (2.53)
If $p_1, p_2, \ldots, p_k$ are the distinct primes dividing n, then $\phi(n)= n(1-1/p_1)(1-1/p_2)\ldots(1-1/p_k)$.

Theorem (2.64)
$\sum_{d|n} \phi(d)=n$.

3

Definition: modular multiplication (3.5)
If the integer a belongs to the residue class $[a]$ modulo n, and the integer b belongs to the residue class $[b]$ modulo n, then multiplication modulo n is defined on $\mathbf{Z}_n$ so that $[a]\times[b]=[ab]$.

Fermat's theorem (3.19)
For any integer x and any prime number p, $x^p \equiv x \pmod{p}$.

Wilson's theorem (3.25)
$(p-1)! \equiv -1 \pmod{p}$, for any prime number p.

Theorem: linear congruences (3.33)
If $d = \gcd(a, n)$, then $ax \equiv b \pmod{n}$ has d solutions which are not congruent modulo n when $d|b$, and no solutions otherwise.

Definition (3.36)
The set of units in $\mathbf{Z}_n$ is denoted by $\mathbf{M}_n$. That is, $\mathbf{M}_n$ consists of those elements of $\mathbf{Z}_n$ which admit an inverse in $(\mathbf{Z}_n, \times)$. An integer from each of the residue classes in $\mathbf{M}_n$ forms a reduced set of residues modulo n.

Theorem (3.36)
$|\mathbf{M}_n| = \phi(n)$.

Fermat–Euler theorem (3.41)
If a and n are coprime, then $a^{\phi(n)} \equiv 1 \pmod{n}$.

Lagrange's theorem on polynomials (3.57)
A polynomial of degree n cannot have more than n distinct zeros. This holds over a field.

Definition: primitive root (3.62)
If $[a]$ is a generator of the group $(\mathbf{M}_n, \times)$, then the integer a is said to be a primitive root modulo n.

Theorem (3.61, 3.65, 3.74)
If n is a power of an odd prime or twice the power of an odd prime, then there exists a primitive root modulo n.

Chevalley's theorem (3.87)
A polynomial in n variables with integral coefficients and of maximum degree less than n, and constant term 0, has a non-trivial solution modulo p.

4

Definition: quadratic residue (4.3, 4.8)
If $[a]$ is a square in the group $(\mathbf{M}_n, \times)$, then a is said to be a quadratic residue modulo n. If $[a]$ is a non-square in the group $(\mathbf{M}_n, \times)$, then a is said to be a quadratic non-residue modulo n.

Definition: Legendre symbol (4.15)
For a prime number p, $\left(\frac{a}{p}\right)$ denotes $+1$ when a is a quadratic residue modulo p, -1 when a is a quadratic non-residue modulo p, and 0 otherwise.

Theorems (4.14, 4.15, 4.41)

$$\left(\frac{a}{p}\right) \equiv a^{\frac{1}{2}(p-1)} \pmod{p},$$

$$\left(\frac{ab}{p}\right) = \left(\frac{a}{p}\right)\left(\frac{b}{p}\right), \left(\frac{2}{p}\right) = (-1)^{\frac{1}{8}(p^2-1)}.$$

Theorem (4.18)
If p is an odd prime, then -1 is a quadratic residue modulo p when $p \equiv 1 \pmod{4}$ and -1 is a quadratic non-residue when $p \equiv 3 \pmod{4}$.

Gauss' lemma (4.43)
If p is an odd prime, and a is an integer which does not have a factor p, then $\left(\frac{a}{p}\right) = (-1)^l$ where l is the number of minus signs

occurring in the list of numerically least residues congruent to

$a \cdot 1, a \cdot 2, \ldots, a \cdot \frac{1}{2}(p-1)$.

Law of quadratic reciprocity (4.62)

If p and q are distinct odd primes, then

$$\left(\frac{p}{q}\right)\left(\frac{q}{p}\right) = (-1)^{\frac{1}{4}(p-1)(q-1)}$$

5

Definition: Pythagorean triple (5.5)

(x, y, z) is called a Pythagorean triple when x, y and z are integers and $x^2+y^2=z^2$. If, moreover, $\gcd(x, y, z)=1$, the triple is said to be primitive.

Theorem: Pythagorean triples (5.15)

If (x, y, z) is a primitive Pythagorean triple, then either x or y is even, and if x is even, then $x=2pq$, $y=p^2-q^2$, $z=p^2+q^2$, for some coprime integers p and q.

Definition: method of descent (n 5.20)

This is the name given to an application of the well-ordering principle devised by Fermat.

Theorem (5.21)

There are no non-zero integers satisfying $x^4+y^4=z^4$.

Theorem (5.68)

There are no non-zero integers satisfying $x^3+y^3=z^3$.

6

Theorem (6.10, 6.11, 6.12)

If p is a prime number, $p|x^2+y^2$, and neither x nor y has a factor p, then $p=2$ or $p\equiv 1 \pmod 4$.

If q is a prime number, $q\equiv 3 \pmod 4$, and $q|x^2+y^2$, then q appears to an even power in the prime factorisation of x^2+y^2.

Theorem (6.15)

$(a^2+b^2)(c^2+d^2)=(ad-bc)^2+(ac+bd)^2$.

Theorem (6.33, 6.36)

Every prime number congruent to 1 (mod 4) can be expressed as the sum of two squares; this can be done in exactly one way.

Theorem (6.34)

A positive integer can be expressed as the sum of two squares if and only if, in its prime factorisation, prime numbers congruent to 3 (mod 4) occur to an even power.

Lagrange's four square theorem (6.40)

$$(a^2+b^2+c^2+d^2)(x^2+y^2+z^2+t^2)$$
$$=(ax-by-cz-dt)^2+(ay+bx+ct-dz)^2$$
$$+(az-bt+cx+dy)^2+(at+bz-cy+dx)^2.$$

Theorem (6.51)
Every prime number can be expressed as the sum of four squares.

Theorem (6.52)
Every positive integer can be expressed as the sum of four squares.

Theorem (6.54)
A positive integer of the form $4^h(8k+7)$ cannot be expressed as the sum of three squares.

7

Definition: partition (7.1)
If n is the sum of a set of positive integers, then the components of that sum are called a partition of n.

Definition: Ferrers' graph (7.2)
The graph of a partition is a display of nodes in which each column displays exactly one component and the columns are arranged from left to right in descending order.

Definition: conjugate partitions (7.3)
Two partitions are said to be conjugate when the graph of one is obtained by transposing the rows and columns of the other.

Definition: generating function (7.16)
The formal power series $f(x)=1+\sum_{n=1}^{\infty} a_n x^n$ is called the generating function of the sequence (a_n).

Definition: $p_m(n)$ (7.22)
The number of partitions of n into parts of size less than or equal to m is denoted by $p_m(n)$.

Theorem (7.22)
The generating function for $(p_m(n))$ is

$$\frac{1}{(1-x)(1-x^2)\dots(1-x^m)}.$$

Euler's theorem (7.47)

$$(1-x)(1-x^2)\dots(1-x^n)\dots=\sum_{n=-\infty}^{\infty}(-1)^n x^{\frac{1}{2}n(3n-1)}.$$

8

Theorem (8.9)

The linear transformation $(x, y) \to (x, y)\begin{pmatrix} a & b \\ c & d \end{pmatrix}$ maps $\mathbf{Z}^2$ onto $\mathbf{Z}^2$ if and only if a, b, c and d are integers and $ad - bc = \pm 1$.

Definition: unimodular transformation (8.15)

A linear transformation of $\mathbf{R}^2$ is said to be unimodular when it maps $\mathbf{Z}^2$ onto $\mathbf{Z}^2$.

Definition: unimodular matrix (8.15)

If $(x, y) \to (x, y)A$ is a unimodular transformation, then A is said to be a unimodular matrix.

Definition: equivalent quadratic forms (8.25)

If $(x, y) \to (px + ry, qx + sy)$ is a unimodular transformation $ax^2 + bxy + cy^2$ and $a(px + ry)^2 + b(px + ry)(qx + sy) + c(qx + sy)^2$ are said to be equivalent quadratic forms.

Theorem (8.25)

Equivalent quadratic forms adopt the same set of values.

Definition: discriminant (8.37)

The discriminant for the quadratic form $ax^2 + bxy + cy^2$ is $b^2 - 4ac$.

Theorem (8.37)

Equivalent quadratic forms have the same discriminant.

Definition: definite forms (8.38)

The quadratic form $ax^2 + bxy + cy^2$ is said to be positive definite when $b^2 - 4ac < 0$ and $a > 0$; negative definite when $b^2 - 4ac < 0$ and $a < 0$; indefinite when $b^2 - 4ac > 0$.

Definition: proper representation (8.48)

The integer n is said to be properly represented by the quadratic form $ax^2 + bxy + cy^2$ when $n = ap^2 + bpq + cq^2$ for $\gcd(p, q) = 1$ or $\{|p|, |q|\} = \{0, 1\}$.

Theorem (8.49, 8.50)

The integer n is properly represented by the form $ax^2 + bxy + cy^2$ if and only if there is an equivalent form in which n is the coefficient of x^2.

Definition (8.57)

The positive definite quadratic form $ax^2 + bxy + cy^2$ is said to be reduced if $0 \leqslant b \leqslant a \leqslant c$.

Theorem: canonical form for equivalence (8.56, 8.66)

Every positive definite quadratic form is equivalent to a unique reduced form.

Definition: automorph (8.73)

An automorph of a quadratic form is a unimodular transformation which maps the quadratic form onto itself.

Theorem (8.77)

The group of automorphs of a positive definite quadratic form is either a cyclic group of order 2 or a dihedral group of order 4, 8 or 12.

9

Definition: fundamental parallelogram (9.5)

If a subgroup G of $\mathbf{Z}^2$ is generated by (a, b) and (c, d), and neither of these ordered pairs is a scalar multiple of the other, then the parallelogram with vertices $(0, 0)$, (a, b), (c, d) and $(a+c, b+d)$ is said to form a fundamental parallelogram for G.

Theorem (9.23)

If G is a subgroup of $\mathbf{Z}^2$ with fundamental parallelogram Π, then the index of G in $\mathbf{Z}^2$ is the area of Π.

Minkowski's theorem (9.39)

An open convex region in $\mathbf{R}^2$ which is symmetrical about $(0, 0)$ and has area greater than 4 contains another point of $\mathbf{Z}^2$.

Definition: fundamental parallelepiped (9.57)

If a subgroup G of $\mathbf{Z}^3$ is generated by $A=(a_1, a_2, a_3)$, $B=(b_1, b_2, b_3)$ and $C=(c_1, c_2, c_3)$, and $aA+bB+cC=(0, 0, 0)$ only when $a=b=c=0$, then the parallelepiped with vertices $(0, 0, 0)$, A, B, C, $B+C$, $C+A$, $A+B$, $A+B+C$ is said to form a fundamental parallelepiped for G.

Theorem (9.66)

If G is a subgroup of $\mathbf{Z}^3$ with fundamental parallelepiped Π, then the index of G in $\mathbf{Z}^3$ is the volume of Π.

Minkowski's theorem (9.71)

An open convex region in $\mathbf{R}^3$ which is symmetrical about $(0, 0, 0)$ and has volume greater than 8 contains another point of $\mathbf{Z}^3$.

Legendre's theorem (9.86)

If a, b and c have no squares in their prime factorisations and are coprime in pairs, then there exists a non-zero integral solution of $ax^2+by^2+cz^2=0$, provided that there exists a solution (x_1, y_1, z_1) of $ax^2+by^2+cz^2 \equiv 0 \pmod{4abc}$ with $\gcd(ax_1, by_1, cz_1, 4abc)=1$.

10

Theorem (10.6)

If a is a non-square rational, then $\sqrt{a}$ is irrational.

Definition (10.10)

$$[a_1, a_2, a_3, \ldots] = a_1 + \cfrac{1}{a_2 + \cfrac{1}{a_3 + \ddots}}$$

The a_i are real numbers, and when $i > 1$, $a_i \geqslant 1$.

Theorem (10.13)

If $[a_1, a_2, a_3, \ldots, a_k] = p_k/q_k$, then for

$$n \geqslant 3, \frac{p_n}{q_n} = \frac{a_n p_{n-1} + p_{n-2}}{a_n q_{n-1} + q_{n-2}}.$$

Definition: convergent (10.18)

If $k \leqslant n$, then p_k/q_k is the kth convergent of $[a_1, a_2, \ldots, a_n]$.

Theorem (10.22)

Each rational number is equal to exactly two distinct continued fractions, one with an odd number of terms and one with an even number of terms.

Theorem (10.23, 10.24, 10.25)

Each irrational number x has a unique expansion as a continued fraction. The sequence of convergents of this continued fraction converges to x. Every infinite continued fraction arises in this way.

Definition: quadratic irrational and conjugate (10.52)

An irrational number which is a solution of a quadratic equation with integral coefficients is called a quadratic irrational. The other solution of the same quadratic equation is called the conjugate irrational.

Theorem (10.37, 10.53)

Every purely periodic continued fraction corresponds to a quadratic irrational greater than 1 whose conjugate is negative and greater than -1, and conversely.

Pell's equation (10.61)

If the period in the continued fraction for $\sqrt{d}$ is of length n, then $p_n^2 - dq_n^2 = (-1)^n$, where p_n/q_n is the nth convergent to $\sqrt{d}$.

Lagrange's theorem on quadratic irrationals (10.76, 10.77)

The continued fraction for every quadratic irrational is ultimately periodic and conversely.

Theorem (10.81, 10.82)

If a and b are coprime positive integers, the automorphs of the quadratic form $ax^2 - by^2$ form an infinite cyclic group.

11

Theorem (11.10, 11.22)

For any real number α and any positive integer n, it is possible to find integers p and q such that $|\alpha - p/q| < 1/q^2$, with $0 < q \leqslant n$.

Hurwitz' theorem (11.31)

If α is an irrational number, then at least one out of every three consecutive convergents to α satisfies $|\alpha - p/q| < 1/\sqrt{5}\, q^2$.

Liouville's theorem (11.42)

If α is an algebraic number of degree $n \geqslant 2$, then there exists a positive real number A such that $|\alpha - p/q| > 1/Aq^n$ for all integers p and all positive integers q.

INDEX

All decimal numbers (2.25 etc.) without the prefix n (for 'note') refer to questions